Technikzukünfte, Wissenschaft und Gesellschaft / Futures of Technology, Science and Society

Herausgegeben von
A. Grunwald, Karlsruhe, Deutschland
R. Heil, Karlsruhe, Deutschland
C. Coenen, Karlsruhe, Deutschland

Diese interdisziplinäre Buchreihe ist Technikzukünften in ihren wissenschaftlichen und gesellschaftlichen Kontexten gewidmet. Der Plural „Zukünfte" ist dabei Programm. Denn erstens wird ein breites Spektrum wissenschaftlich-technischer Entwicklungen beleuchtet, und zweitens sind Debatten zu Technowissenschaften wie u.a. den Bio-, Informations-, Nano- und Neurotechnologien oder der Robotik durch eine Vielzahl von Perspektiven und Interessen bestimmt. Diese Zukünfte beeinflussen einerseits den Verlauf des Fortschritts, seine Ergebnisse und Folgen, z. B. durch Ausgestaltung der wissenschaftlichen Agenda. Andererseits sind wissenschaftlich-technische Neuerungen Anlass, neue Zukünfte mit anderen gesellschaftlichen Implikationen auszudenken. Diese Wechselseitigkeit reflektierend, befasst sich die Reihe vorrangig mit der sozialen und kulturellen Prägung von Naturwissenschaft und Technik, der verantwortlichen Gestaltung ihrer Ergebnisse in der Gesellschaft sowie mit den Auswirkungen auf unsere Bilder vom Menschen.

This interdisciplinary series of books is devoted to technology futures in their scientific and societal contexts. The use of the plural "futures" is by no means accidental: firstly, light is to be shed on a broad spectrum of developments in science and technology; secondly, debates on technoscientific fields such as biotechnology, information technology, nanotechnology, neurotechnology and robotics are influenced by a multitude of viewpoints and interests. On the one hand, these futures have an impact on the way advances are made, as well as on their results and consequences, for example by shaping the scientific agenda. On the other hand, scientific and technological innovations offer an opportunity to conceive of new futures with different implications for society. Reflecting this reciprocity, the series concentrates primarily on the way in which science and technology are influenced social and culturally, on how their results can be shaped in a responsible manner in society, and on the way they affect our images of humankind.

Prof. Dr. Armin Grunwald, Physiker, Mathematiker und Philosoph, lehrt Technikphilosophie und Technikethik am Karlsruher Institut für Technologie (KIT), ist Leiter des Instituts für Technikfolgenabschätzung und Systemanalyse (ITAS) in Karlsruhe und Leiter des Büros für Technikfolgen-Abschätzung beim Deutschen Bundestag (TAB) in Berlin. / Professor Armin Grunwald, physicist, mathematician and philosopher, teaches the philosophy and ethics of technology at the Karlsruhe Institute of Technology (KIT), and is the director of the Institute for Technology Assessment and Systems Analysis (ITAS) in Karlsruhe and of the Office of Technology Assessment at the German Bundestag (TAB) in Berlin.

Reinhard Heil, Philosoph, ist wissenschaftlicher Mitarbeiter am KIT-ITAS. / Reinhard Heil, philosopher, is a researcher at KIT-ITAS.

Christopher Coenen, Politikwissenschaftler, ist wissenschaftlicher Mitarbeiter am KIT-ITAS und Herausgeberder Zeitschrift ‚NanoEthics: Studies of New and Emerging Technologies'. / Christopher Coenen, political scientist, is a researcher at KIT-ITAS and the editor-in-chief of the journal 'NanoEthics: Studies of New and Emerging Technologies.'

Bettina-Johanna Krings
Hannot Rodríguez · Anna Schleisiek
Editors

Scientific Knowledge
and the Transgression
of Boundaries

Springer VS

Editors
Bettina-Johanna Krings
Karlsruher Institut für Technologie (KIT)
Karlsruhe, Germany

Anna Schleisiek
Berlin, Germany

Hannot Rodríguez
University of the Basque Country
UPV/EHU
Vitoria-Gasteiz, Spain

Technikzukünfte, Wissenschaft und Gesellschaft / Futures of Technology,
Science and Society
ISBN 978-3-658-14448-7   ISBN 978-3-658-14449-4   (eBook)
DOI 10.1007/978-3-658-14449-4

Library of Congress Control Number: 2016946002

Die Deutsche Nationalbibliothek verzeichnet diese Publikation in der Deutschen Nationalbibliografie; detaillierte bibliografische Daten sind im Internet über http://dnb.d-nb.de abrufbar.

Springer VS
© Springer Fachmedien Wiesbaden GmbH 2016

Lektorat: Frank Schindler

Gedruckt auf säurefreiem und chlorfrei gebleichtem Papier

Springer VS ist Teil von Springer Nature
Die eingetragene Gesellschaft ist Springer Fachmedien Wiesbaden GmbH
Die Anschrift der Gesellschaft ist: Abraham-Lincoln-Str. 46, 65189 Wiesbaden, Germany

Preface

Philosophy and sociology of science have historically tended to think of science as an autonomous system strictly isolated from the rest of society. Also, the boundaries between different scientific disciplines have traditionally been conceived as permanent and distinct. But, on the contrary, science seems to be an activity that constantly transgresses, erases, and rebuilds disciplinary and societal boundaries.

While previous boundaries of scientific knowledge are overcome, progress in knowledge reshapes or creates new boundaries both within and outside the particular discipline concerned. These redefinitions of boundaries have marked the entire historical process of scientific development, challenging the capacity of the disciplines involved to come to terms with new forms of knowledge production. Processes of transgression and recreation of new boundaries, though, cannot be conceived as a mere by-product of scientific progress: Current research policies explicitly aim to promote cross-disciplinary R&D projects such as Nano-Bio-Info-Cogno convergence.

However, scientific developments are not only seen as milestones of progress: In many cases they trigger heated debates in society due to their ambivalent and complex effects on social processes. Research in fields such as the life sciences, emerging technologies or brain research has given rise to social doubts and uncertainties. They again question the boundaries between humans and nature and force us to reexamine the role of science and technology in the dynamics of social progress. As a result, there is an increasing demand for the development and application of codes of ethics in these fields and for socially "responsible innovation", allowing and requiring interdisciplinary collaboration between social and natural scientists and the participation of stakeholder groups, including the public, in R&D activities.

The aim of this book is to provide a thorough and comprehensive analysis of the dynamics of the transgression of social–scientific boundaries. It contains ten contributions, which provide significant theoretical and empirical insights into disciplinary transgressions. All contributions are based on presentations made at the International Graduate Summer School "Scientific Knowledge and the Transgression of Boundaries", held at Donostia-San Sebastián in 2012 (from July 28 to August 1). The Summer School was part of the XXXI Summer Courses and XXIV European Courses of the University of the Basque Country (UPV/EHU), and was co-organized by the Institute for Technology Assessment and Systems Analysis (ITAS) at Karlsruhe Institute of Technology (KIT) and the UPV/EHU. This event was the first of a series of International Graduate Summer Schools organized biannually by the University of the Basque Country and the Karlsruhe Institute of Technology on a regular cooperation basis between the two institutions.

Karlsruhe, Germany   Bettina-Johanna Krings
Vitoria-Gasteiz, Spain   Hannot Rodríguez
Berlin, Germany   Anna Schleisiek

Acknowledgments

First, we would like to express our gratitude to the Institute for Technology Assessment and Systems Analysis (ITAS) at Karlsruhe Institute of Technology (KIT) and the University of the Basque Country (UPV/EHU) for, respectively, funding and hosting the 5-day International Graduate Summer School on which this book is based. In particular, we want to acknowledge Prof. Gotthard Bechmann (ITAS-KIT) and Prof. Andoni Ibarra (UPV/EHU) for establishing the basis for such interinstitutional partnership, as well as the administrative personnel both at ITAS-KIT and the UPV/EHU for their highly professional and attentive support during the whole organization process.

Second, our sincere thanks go to the authors for their contributions to the volume and the excellent cooperation during the process.

Third, we would like to thank Sylke Wintzer (ITAS-KIT) for editing and proofreading the articles—most of them written by non-native English speakers—as well as for preparing the layout of the book.

Fourth, we are grateful to Armin Grunwald, Reinhard Heil and Christopher Coenen (all ITAS-KIT), the editors of the "Futures of Technology, Science and Society" Springer VS Series, for kindly inviting us to contribute to it.

Last but not least, Hannot Rodríguez would like to thank the Karlsruhe Institute of Technology (KIT) for awarding him a fellowship in 2012 to conduct research at the Institute for Technology Assessment and Systems Analysis (ITAS), which allowed him to participate in the organization of the summer school on which this publication is based. He also wants to acknowledge the support of the Spanish Ministry of Economy and Competitiveness under grants FFI2011-24414 and

FFI2012-33550, the Basque Government's Department of Education, Language Policy and Culture under grant IT644-13, and the University of the Basque Country UPV/EHU under grant EHUA15/13.

Bettina-Johanna Krings
Hannot Rodríguez
Anna Schleisiek

Contents

Part III Radical Transgression of Boundaries

Editors and Contributors

About the Editors

Bettina-Johanna Krings Ph.D. in sociology, M.A. in sociology, political science, and anthropology. Since 1994, research fellow at the Institute for Technology Assessment and Systems Analysis (ITAS) at the Karlsruhe Institute of Technology (KIT). Current position: Head of research department "Knowledge Society and Knowledge Policy" at ITAS. Since 1998, she has developed the research focus on "Technology and its impact on work structures" at ITAS where she has led many projects and research activities. In addition to her research on these topics, she is working on the methods and theory of technology assessment and on gender theory.

Hannot Rodríguez Born in 1976. B.A. in philosophy, M.A. in philosophy and history of science and technology, and Ph.D. in philosophy (2007) from the University of the Basque Country UPV/EHU. He is Assistant Professor at the Department of Philosophy of the UPV/EHU. He is also a member of the Miguel Sánchez-Mazas Chair at UPV/EHU, and an affiliate of the Consortium for Science, Policy & Outcomes (CSPO) at Arizona State University. His research interests include risk governance of emerging technologies, sociotechnical integration in science and technology policy, responsible innovation, and philosophical and social dimensions of science and technology.

Anna Schleisiek (Dipl.-Soz.) She studied sociology, political science, and statistics at the Freie Universität in Berlin. She has been a researcher at the Institute for Technology Assessment and Systems Analysis (ITAS) at the Karlsruhe Institute of Technology (KIT) since 2006. There, her previous work was focused on the practices of knowledge and technology transfer. Currently, she is doing research on the role of economic principles in the scientific practice of research teams in materials research. Other research interests include the sociology of science, the sociology of Pierre Bourdieu, and transformative processes in science and society.

Contributors

Nuno Boavida He graduated in industrial production engineering at the Universidade Nova de Lisboa in 1999. In 2001, he finished a M.Sc. in industrial relations and personnel management at the London School of Economics, where he focused his thesis on the Portuguese labor movement. From 2003 to 2011, he received two scholarships from the Portuguese National Science Foundation to work at the Observatory for Science, Technology, and Higher Education. He is currently a Ph.D. candidate at the Universidade Nova de Lisboa, writing a thesis about the role of indicators and evidences in decisions regarding technology innovation. Since 2012, he has been an invited researcher at the Institute for Technology Assessment and Systems Analysis (ITAS) of the Karlsruhe Institute of Technology (KIT). His research interests focus on technology assessment, the use of indicators and evidences, decision-making in technology, and innovation policy.

Kathrin Friedrich She works as a research associate in the project "Image Guidance" at the DFG Cluster of Excellence "Image Knowledge Gestaltung", Humboldt University Berlin. In October 2014, she was a guest researcher at the Department for Thematic Studies, Linköping University. Previously, she worked at the Academy of Media Arts Cologne and in the collaborative BMBF research project "Embodied Information. 'Lifelike' Algorithms and Cellular 'Machines'". She studied media studies, law, and sociology at the University of Marburg. Research areas: Image-guided interventions in medicine, warfare and architecture, medical imaging, computer-aided design and serious gaming in biology, media theory, and software studies. Selected publications: "Achromatic reasoning—On the relation of gray and scale in radiology", *Leonardo* 48(1)/2015: 66–67; "Automated killing and mediated caring. How image-guided robotic intervention redefines radiosurgical practice", *Proceedings of the 50th Anniversary Convention of the AISB 2014* (with M. Queisner); "Digital 'faces' of synthetic biology", *Studies in History and Philosophy of Science Part C* 44/2013: 217–224; "'Sehkollektiv': Sight styles in diagnostic computed tomography", *Medicine Studies* 2/2010:185–195.

Steve Fuller Born in 1959. Ph.D. University of Pittsburgh (history and philosophy of science). Auguste Comte Professor of Social Epistemology in the Department of Sociology, University of Warwick, UK. Founder of the first journal and author of the first book on "social epistemology" and author of 21 books plus many articles, including translations into more than 20 languages. Recently, he completed a trilogy relating to the idea of a "post-" or "trans-" human future, all published by Palgrave Macmillan: *Humanity 2.0: What It Means to Be Human Past, Present and Future* (2011), *Preparing for Life in Humanity 2.0* (2012), and (with Veronika Lipinska) *The Proactionary Imperative: A Foundation for Transhumanism* (2014). His latest book is *Knowledge: The Philosophical Quest in History* (Routledge, 2015). He was awarded a D.Litt. by the University of Warwick in 2007 for sustained lifelong contributions to scholarship.

Genco Guralp He is a doctoral candidate in philosophy at Johns Hopkins University, where he currently holds a fellowship at the Center for Advanced Media Studies. His research interests include historical epistemology, philosophy of science, and science and technology studies. His dissertation examines the discovery of the acceleration of the universe by two independent research teams, using this as a test case to probe the debate between the Bayesian and the error statistical philosophies of science. A recent paper in which he studied the early attempts to measure the Hubble Constant, "Calibrating the Universe: The Beginning and End of the Hubble Wars," was published in: O. Schlaudt and L. Huber (eds.) *Standardization in Measurement: Philosophical, Historical and Sociological Issues* (Pickering & Chatto Publishers, 2015).

Judith Igelsböck Born in 1983. Ph.D. candidate in science and technology studies, University of Vienna. Holds a master's degree in sociology. Worked at the Center for Advanced Studies and Research in Information and Communication Technologies and Society at the University of Salzburg and the Department of Science and Technology Studies at the University of Vienna. Interested in the following research areas: Knowing in the "digital age", new ways of knowledge production, social innovation, changing science–society relations, controversy mapping, digital humanities, and sociotechnical assemblages. Her thesis, entitled "The performance and transformation of evidence in transdisciplinary research contexts", takes into view attempts to open up science to various forms of expertise to make it become more responsive to contemporary societal problems. She focuses on evidence to show how roles and responsibilities are re-scribed and redistributed between science and society, but also how standards, long-standing myths, and frequently rehearsed procedures are limiting the scope for new entanglements between science and society.

Jaume Navarro He is Ikerbasque Research Professor at the University of the Basque Country UPV/EHU. He trained in physics, philosophy, and the history of science and has an international research record having spent several years at the University of Cambridge, Imperial College London, the Max Planck Institute for the History of Science, and the University of Exeter. He is author, among other books, of *A History of the Electron. J.J. and G.P. Thomson* (Cambridge University Press, 2012).

Fabrice Pataut He has been a Lavoisier and a von Humboldt scholar. He is a researcher at the *Centre National de la Recherche Scientifique*. He has published numerous articles, both in French and English, on the philosophy of language, philosophy of logic, and philosophy of mathematics, as well as occasional essays on ethics. He is the guest editor, with Daniele Molinini and Andrea Sereni, of a forthcoming issue of *Synthese* on "Indispensability and Explanation", and the editor of the forthcoming book *Truth, Objects, Infinity: New Perspectives on the Philosophy of Paul Benacerraf*, to be published by Springer-Verlag.

Farah Purwaningrum She is a sociologist with an interdisciplinary background in law. She holds law degrees from Universitas Islam Indonesia, Yogyakarta and the London School of Economics and Political Science, UK. She completed her Dr. Phil. at the Rheinische Friedrich-Wilhelms-Universität Bonn, Germany in 2012. She is currently based at the Institute of Asian Studies at Universiti Brunei Darussalam, at which she holds a lectureship in sociology. Several of her recent publications include the following: *Knowledge Governance in an Industrial Cluster. The Collaboration between Academia-Industry-Government in Indonesia* (LIT Verlag, 2014), *Towards a Formation of Skillful Worker or Semi Profession? The Implementation of Japanese Flexible Specialisation in the Indonesian Automotive Sector* (paper presented at RSA Winter Conference in London, November 27–28, 2014). She has a keen interest to do research in areas of science policy, science system, and knowledge governance.

Hans-Jörg Rheinberger Born in 1946. He studied philosophy and biology in Tübingen and Berlin, Germany. He received his M.A. in philosophy in 1973, his Ph.D. in biology in 1982, and his habilitation in molecular biology in 1987. He was Assistant Professor at the University of Lübeck, Germany, and Associate Professor at the University of Salzburg, Austria. From 1997 to 2014, he was a scientific member of the Max Planck Society and Director at the Max Planck Institute for the History of Science in Berlin, where he now pursues his research as Director Emeritus.

Pedro Sáez Williams Born in 1981. He is a final year Doctoral Researcher in the Department of Sociology at the University of Warwick. Originally trained as a lawyer, Sáez Williams has been involved in political activism, political work, and legal practice before deciding to engage in the academic world. He holds a LL.B. from *The Universidad Iberoamericana* in Mexico City, an M.A. in social and cultural anthropology from the University of Vienna and an M.Phil. in sociology from the University of Cambridge. His research is concerned with the dimensionality of normativity and social authority (both cognitive and political), specifically focused on developing an account of normativity compatible with naturalism.

Introduction: Possibilities and Limits of Science-Based Boundary Transgressions

Bettina-Johanna Krings, Hannot Rodríguez
and Anna Schleisiek

The modern scientific and industrial revolutions have provided humanity with a hitherto unknown capacity to understand and transform the world. Natural boundaries and possibilities have been radically redefined by scientific and technological developments. These developments have altered the living conditions of humanity in a fundamental way. For example, one's biological fate is fought and altered by medical advances, which are at the same time progressively modifying the basis of genomic research (Rabinow 1996; Fukuyama 2002). Furthermore, the lifespan of food has been dramatically prolonged due to the use of chemicals and biotech solutions, at the same time that food maturation time has become progressively shorter. In fact, we can observe processes speeding up everywhere in our daily life. This is not only true of organic processes, but also, for example, with regard to mobility—airplanes, for instance, make it possible for us to fly in spite of our "terrestrial" biology. In other words, new realms of possibilities for human action and capabilities come from scientific and technological progress, which is the epitome of modern times (Ridley 2010; Daston and Galison 2007).

Given their transformative capacity, science and technology have become key innovation agents for industrial and economic progress in highly industrialized

B.-J. Krings (✉)
Karlsruhe, Germany
e-mail: bettina-johanna.krings@kit.edu

H. Rodríguez
Vitoria-Gasteiz, Spain
e-mail: hannot.rodriguez@ehu.eus

A. Schleisiek
Berlin, Germany
e-mail: anna.schleisiek@yahoo.de

© Springer Fachmedien Wiesbaden GmbH 2016
B.-J. Krings et al. (eds.), *Scientific Knowledge and the Transgression of Boundaries,* Technikzukünfte, Wissenschaft und Gesellschaft / Futures of Technology, Science and Society, DOI 10.1007/978-3-658-14449-4_1

societies (Marklund et al. 2009). In the opinion of the European Commission (EC), "Research and innovation [...] improve Europe's competitiveness, boost growth and create jobs" (EC 2014a, p. 3).[1] Similarly, the United States (US) National Science Foundation (NSF) claims that "Investing in S&E [science and engineering] is widely recognized as an essential pathway to the nation's future prosperity" (NSF 2011, p. 1), and non-Western governments have also shown an increasing interest in innovation in the last decades, especially in Asia (Kondo et al. 2006; Parayil and D'Costa 2009). The socio-economic function of science and technology asserted in these statements is strongly embedded in political socio-technical assumptions, or "imaginaries", in which future visions of societies are projected and constituted on the basis of business-oriented technological innovation, and economic growth tends to be assumed to be an almost sufficient condition for societal well-being (Guston et al. 2014; Jasanoff 2004).

However, in addition to the general rhetoric on business and socio-economic progress, support of science and technology is also justified by their capacity to solve complex socio-environmental challenges. For instance, in the opinion of the EC:

> Smart investment, notably in research and innovation, is vital in order to maintain high standards of living while dealing with pressing societal challenges such as climate change, an ageing population, or the move towards a more resource-efficient society (EC 2011, p. 2).

This problem- or challenge-based approach to research and development (R&D) arguably implies at the same time a further strengthening of the problematic nature of traditional disciplinary boundaries in science, resulting from the circumstance that research here is intended to produce knowledge about very complex and heterogeneous problems despite being conducted in well-defined disciplinary subjects (Bruce et al. 2004, pp. 457–459). As claimed, again, by the EC:

> A challenge-based approach will bring together resources and knowledge across different fields, technologies and disciplines, including social sciences and the humanities (EC 2011, p. 5).

---

[1] Upstream policy rhetoric on competitiveness and the strategic value of innovation is reproduced for instance by the European Special Interest Group (SIG) on Cooperative Robotics's white paper, where SIG's constitution and role are justified in terms of meeting "the tremendously increased interest in cooperative robots of different types for many emerging applications, and to foster Europe's position as leader in the field" (Saffiotti and Lima 2008, p. 4).

This turn toward more heterogeneous and less disciplinary forms of research practice can also be directed under certain circumstances not just by academic but also extra-academic actors (more about which below).

Nevertheless, calls for us to rethink science's structure, role and goals do not just occur in contexts where science is conceived as the prime "solving agent" of a given problem, but also in relation to research itself (Bechmann et al. 2007). In other words, scientific and technological developments are not only appraised in our societies as milestones of socio-economic progress and societal well-being, but also as sources of serious concerns, exposing these developments as a result to constant interpretation and evaluation, even hostility, with regard to their (i) ecological and health, (ii) socio-ethical, and (iii) broader cultural impacts:

(i) Science and technology are a matter of environmental and human health concern when they either are used intentionally to provoke damage—i.e., military-purposed innovation (Black 2013)[2]—or cause unintended side effects such as technological accidents (Perrow 1984) or cumulative environmental contamination (Giddens 2009). Altogether, this impact arguably contributes to the dramatic modification of the Earth's ecosystem (Millenium Ecosystem Development 2005), a topic that has been raised since the early 1970s when issues such as the emergence of a global ecological crisis started being debated as posing a limit to technoeconomic growth (Meadows et al. 1972). The matters at stake here are diverse, but several of them that these issues point to are the moral and political responsibilities of scientists and engineers, the systemic limits of knowledge, and the arguably exaggerated belief in human control capabilities.

(ii) Together with the ecological crisis, socio-ethical issues and insecurity pervade the way in which the future evolution of our societies is understood and valued. For instance, developments in modern medicine extend our lifetime but at the cost of raising side effects like new socio-ethical concerns about longevity's impact on environmental sustainability (Wright and Lund 2000; Pillemer 2011) or the viability of public pension and health care systems (Watts-Roy and Williamson 2009; Crystal and Siegel 2009). Thus, technical progress in medicine opens up debates about fundamental socio-ethical issues, combined with a lack of shared normative criteria and commitments (Sandel 2007). Another main issue that falls into this category of concern has to do with the systematic unequal societal

---

[2]There is probably no episode that better illustrates the destructive capacity of science-based technical artifacts than the dropping of atomic bombs over Hiroshima and Nagasaki in August 1945 during the final stages of World War II. In total, more than 300,000 people (most of them civilians) were killed and tens of thousands more were injured as a consequence of the atomic bombings (Cheek 2005).

distribution of scientific-technological innovations, guided more toward the satisfaction of industrial interests and superfluous, environmentally damaging consumerism than to satisfying basic human needs and aspirations (Cozzens and Wetmore 2010; Sarewitz 1996, pp. 117–140).

(iii) Last but not least, progress in science and technology provokes a constant and profound transformation of our sense of being human—concerning its nature, capabilities, and limits (Allenby and Sarewitz 2011; Fuller 2011). In fact, scientific knowledge does not just provide accurate theoretical representations of nature, but leads to radical changes in the very meaning of life. For instance, the theories of heliocentric astronomy and natural selection did not just transform the scientific understanding of the solar system and organic life, respectively. They also caused cultural and psychological revolutions, in the sense that they contributed to transforming humanity's self-image and to lowering its "cosmic arrogance", so to speak: Heliocentrism displaced the Earth (and, hence, humanity) from the center of the solar system, and Darwinism represented life as a contingent and evolving reality, in virtue of which humans were a species among others, part of the same biological "tree of life", and not a special being created in the image of God (Dennett 1995; Weinert 2009). In addition to their coincidental implications, scientific arguments are also intentionally—and increasingly—used to rebut nonscientific, or "irrational", worldviews and explanations such as religious ones (Dawkins 2008; Dennett 2006; Krauss 2012), which, in contrast, could arguably be interpreted as being legitimate—i.e., existentially meaningful—expressions of humanity's hermeneutic dimension (Berlin 1990; O'Hear 1996).

Thus, it can be argued that it is precisely by virtue of their pervasiveness and success that science and technology have themselves become more "vulnerable" to contestation and change. In other words, their transformative or constitutive capacity—i.e., their ability to open up new boundaries and possibilities—provokes the emergence of new visions about their capabilities, roles, and goals, which directly influence how they are understood, organized, and practiced. The boundaries of science and technology are therefore also not given; rather, they are constantly challenged and contested (Bauer 1995; Nowotny 2005; Callon et al. 2009).

At the same time, though, science tends to be characterized as a value-free activity, supposedly insulated from societal or ideological considerations.[3]

---

[3]The influence of non-epistemological values on scientific activity has been typically theorized as being limited to factors that are external to the core scientific activity, such as the methodological limits (e.g., ethical limits in experimentation) or the societal and environmental consequences of research (e.g., Rescher 1999, pp. 151–167).

Frequently, socio-environmentally sensitive decisions regarding techno-industrial advances are justified by this characterization. For instance, regulations of technological safety are claimed to be based on sound or objective science—which also implies that alternative socio-technical safety scenarios are systematically discredited in the name of that very objectivity (Rodríguez 2016). In that sense, it can arguably be said that scientific knowledge tends to be represented here as a completely *confined* practice with almost *unconfined* powers and normative legitimacy.

However, close analysis of regulatory science—i.e., the science used to assess the safety and efficiency of technologies to be regulated—shows that non-epistemological considerations necessarily influence how evidence of risk is interpreted by science. This is because our knowledge about risks often remains uncertain and the consequences of accepting or rejecting scientific hypotheses are non-epistemological (e.g., socio-economic or environmental), meaning that, as such, decisions on uncertain risks are invariably influenced by non-epistemological considerations (Douglas 2000). Take, for example, the issue of methodological choice in the field of radiation risk assessment, where basically either of two alternative dose-response models can be used: The quadratic model or the linear model. According to the quadratic model, there is a certain dose threshold below which radiation does not have any significant effect on human health. In contrast, the linear model understands the dose-response relationship in gradual terms, meaning that a constant exposure to low doses of radiation is believed to have important negative effects on humans (Longino 1990, pp. 391–397). Nevertheless, due to great scientific uncertainty about the biological risks of low-level radiation—as a consequence, in part, of severe statistical-temporal limitations—the choice of model is inevitably made on basis of non-epistemological (i.e., socio-pragmatic) considerations such as "beliefs in the social utility of nuclear energy programs or other nuclear technologies, skepticism regarding their value, or interest in competing concerns such as health" (Longino 1990, p. 398).

One of the implications of accepting the fact that social values or interests play a fundamental role in the constitution of scientific knowledge is that science is seriously undermined when it is utilized as an objective limit to constrain debates on techno-industrial progress. For instance, former European Commissioner for Health and Consumer Protection (1999–2004) David Byrne attempted to delegitimize the public backlash against agro-food biotechnology—based to a large extent on safety concerns (Gaskell 2008)—by claiming that "we need to get away from the emotional, the irrational and the bullying tactics".[4] However, as seen above, it seems problematic to argue that science is completely insulated from societal—i.e., "impure"—influences. Rather, it is arguably more correct to approach controversies about the magnitude of technological risks as reflecting a divergence over the ideologically informed criteria used to interpret incomplete evidence, provided that in regulatory contexts "science and policy are difficult to distinguish" (Jasanoff 1990, p. 79).

In fact, crises such as the transgenic agro-food backlash and others that took place in Europe during the 1990s (e.g., mad cows, dioxin-contaminated chicken, foot-and-mouth disease), which were seen by broad sectors of society as the outcomes of industry-oriented and publicly insensitive policies which "undermined public confidence in expert-based policy-making" (EC 2001, p. 19), have brought European Union (EU) policymakers to realize that the innovation system needs to be modified to include broader socio-ethical issues and criteria by which to conduct and evaluate R&D practices. For instance, in the wake of the worst of the societal resistance against agro-food biotechnology, at the beginning of the previous decade, former European Commissioner for Research (1999–2004) Philippe Busquin claimed that

> [...] democratic governance must ensure that social and economic issues are taken into consideration in research activities, and that citizens are informed about and are aware of the social aspects with regard to scientific and technological progress (Busquin 2003, p. 6).

Policy narratives on integrating broader aspects and actors into R&D activities have progressively become more radical, to the point that the EC currently claims to support that

> [...] all societal actors (researchers, citizens, policy makers, business, third sector organisations etc.) [...] work together during the whole research and innovation process in order to better align both the process and its outcomes with the values, needs and expectations of European society (EC 2014b, p. 4).[5]

In other words, citizens here should not just be informed but actively involved in R&D activities. As claimed for instance by the Head of Unit for Nano and Converging Sciences and Technologies of the EC, Christos Tokamanis, attempts have

---

[4]EC, CORDIS (2001) Byrne calls for more realistic approach to GMO risk (11/27/2001). http://cordis.europa.eu/news/rcn/17678_en.html (Accessed 29 Nov 2014).

[5]This approach has been termed "responsible research and innovation" (RRI) (EC 2014b, p. 4).

to be made "to engage citizens as early as possible in all developments and processes" (Tokamanis 2011, p. 10).[6]

Such claims should not be uncritically accepted, though, as evidence demonstrates that actual policy requests for integrating societal and ethical issues and perspectives—including those of the lay public—into European core R&D practices are relatively marginal in comparison to requests for integrating industrial actors and economic considerations (Rodríguez et al. 2013). However, it also does not mean that no significant attempts have been made in order to "modulate" science and engineering R&D activities socially and ethically, i.e., reflexively and critically, as have been reported in US and European contexts, mostly with regard to the effects of integrating experts from the social sciences and humanities into labs (e.g., Fisher 2007; Flipse et al. 2013; Schuurbiers 2011). In any case, it seems reasonable to expect high systemic difficulties and resistance when implementing policy demands for substantial transformations of industry- and profit-oriented R&D innovation systems on the basis of socio-ethical research and criteria. For instance, US legislation on nanotechnology has been reported to be open to interpretation, being viewed both as a call to radically transform the nano innovation system through the integration of societal research and considerations, and as a secondary complement to strategic, profit-driven nano R&D—hence, as a "contradictory intent" (Fisher and Mahajan 2006).

In any case, there is an increasing emergence of alternative "socio-technical imaginaries" beyond the articulated or institutionalized integration fora (Jasanoff and Kim 2009), which come to dispute profit-oriented unidimensional relations between science, technology, and politics. Radical counterproposals such as "keeping technologies out" (Felt 2013) are usually the outcome of a "bottom-up formation process" (Felt 2013, p. 15) constituted around culturally sensitive "technopolitical identities" (Felt 2013, p. 16). In fact, it seems justified to argue that the dominant motivation behind the development of socially more responsible innovation policies has to do with facilitating the societal uptake—and, therefore, viability—of technological innovations in the context of a global, highly competitive knowledge-based economy (Krings 2011). For example, the European regulatory framework for agro-food biotechnology became progressively tighter in

---

[6]Arguably, those institutional demands for democratizing innovation processes have also been influenced by descriptive and normative scholarly theories such as "post-normal science" (Funtowicz and Ravetz 1993), "mode 2 knowledge production" (Gibbons et al. 1994), "constructive technology assessment (CTA)" (Schot and Rip 1997), "real-time technology assessment" (Guston and Sarewitz 2002), or "user innovation" (von Hippel 2005).

parallel to increasing societal resistance against the technology (Macnaghten 2008).[7] As claimed by the EC:

> For Europe to become the most advanced knowledge society in the world, it is imperative that legitimate societal concerns and needs concerning science and technology development are taken on board (EC 2007, p. 4).

Science-based boundary transgression hence represents the potential or possibility for transformation and evolution, but also at the same time the constraints underlying certain non-transgressable boundaries. This is not an exclusive trait of innovation-oriented science, where the strong political commitment toward innovation and its economic impact limits the extent to which more reflexive research practices involving collaboration between natural and social scientists or even between experts and laymen can be implemented and regularized (Rodríguez et al. 2013). On the contrary, interdisciplinary research in less application-oriented, or less strategic, research areas and contexts has to confront the dominant discipline-based research culture in academia, which discourages professionals from pursuing transgressive research paths and careers, for instance, by compelling one to publish in highly ranked journals, where research success is measured according to disciplinary criteria (Bruce et al. 2004, p. 464; Rhoten 2004; van Rijnsoever and Hessels 2011, pp. 469–470).[8] However, the lack of cross-disciplinary collaboration cannot be explained in political or institutional terms alone. Methodological and terminological differences can pose difficult

---

[7]It might be noted, though, that the US, for instance, has also promoted responsible R&D—e.g., US National Science and Technology Council (NSTC) claims "responsible development" of nanotechnology through "engagement with universities, industry, government agencies […], nongovernmental organizations, and other communities" (NSTC 2014, p. 6)—without having suffered such a recent fierce controversy—even if it also cannot be ignored that the country has been the scene of harsh public debate concerning other technologies such as nuclear power (mostly in the 1970s and 1980s) (Nelkin 1994) or the more recent fracking technique (Boudet et al. 2014). In any case, it seems prudent to complement instrumental explanations with normative and substantive considerations (both in the EU and the US), which means that there is some belief that more open and reflexive (i.e., more responsible) research is inherently desirable, and that it leads to better decisions and outcomes, respectively (Stirling 2008).

[8]That is why inter- or transdisciplinary research like research on sustainability is not fully recognized as a research "discipline." As already said, methods of evaluation are usually focused on "classical" performance criteria such as the number of published articles in highly ranked journals, but the objectives of interdisciplinary and transdisciplinary research are more related to the transformative character of problem-solving strategies in public contexts (Grunwald and Schmidt 2005).

obstacles to fruitful collaboration between disciplines (Bruce et al. 2004, pp. 464, 467–468).

Thus, the dynamics and potential for transgression need to be analyzed and understood in terms of the relationships between a complex and heterogeneous set of epistemological, political, economic, and socio-cultural factors. The configuration of the socio-technical network is flexible, though the extent to which it can be transformed depends on how the elements by which it is constituted are valued and interpreted in light of a variety of historical trends, evolutions, and revolutions, as well as of the contingencies and inertias underlying progress and resistance to it. The contributions in this book aim precisely to shed some light on the complexity of transgressive dynamics.

The book has been divided into three main parts, according to three different levels of boundary transgressions. In the Part I, "Disciplinary Transgression of Boundaries", the papers are mainly focused on issues concerning the flexibility and partial distortion of disciplinary methodologies, expertise, roles, and statuses. In Part II, "Extra-disciplinary Transgression of Boundaries", the contributions analyze the relationships—articulated or unarticulated, legitimate or illegitimate—between scientific disciplines and nonscientific actors and issues. Finally, in Part III, "Radical Transgression of Boundaries", the contributions point to more transgressive interpretations of transgression, where the very disciplinary divisions of knowledge are viewed problematically and "boundless" transgression is deemed to be a constitutive characteristic of scientific knowledge.

Part I: Disciplinary Transgression of Boundaries

In the first chapter of the book, "Science and Experiment", Hans-Jörg Rheinberger analyzes the heuristic potential of serendipity and chance in experimental science. These "non-methodological" factors, usually excluded from scientific publications as well as from sociological and philosophical analyses and reconstructions, are claimed by the author to have played a very important role in the history of modern science. In contrast to the productive role of mishaps, or unexpected experimental results, extreme methodological precision and experimental virtuosity are approached as potentially counterproductive research strategies in that they do not leave room for chance and the emergence of potentially productive anomalies. The author relates chance-allowing practices and hence epistemological fruitfulness with the cross-disciplinary scientific culture (which is illustrated by the example of mid-twentieth century molecular biology), which is in contrast to strongly discipline-based experimental practice, where the emergence and exploration of new "epistemic things" is constrained or limited by rigid methodological boundaries.

Next, in her contribution "From 'Imaging 2.0' to 'Imaging 3.0': On the Crisis of Radiology and Its 'Culture Shifts'", Kathrin Friedrich describes ostensively how the transformation of technical equipment in radiology towards digital workflows does not only change the expertise of the radiologists but also work routines and the use of radiology within clinical environments. In that vein, Friedrich aims to answer the following question: "How are technological innovations mutually challenging disciplinary paradigms and specialties as well as the (self-)conception of a certain expertise?" She then argues that innovation-based disciplinary and socio-cultural transformations in clinical contexts are being partially resisted by radiologists. The strong influence of image-guided therapies on medical treatments is claimed to demand an improved dialogue between radiologists and software programmers. Radiology is thus characterized as facing a whole cultural shift involving a transition from diagnosis to therapy, by virtue of which disciplinary boundaries are inevitably being made problematic.

The last paper of this first part is Nuno Boavida's "The Use and Influence of Indicators in Decisions about Technology Innovation: Quantitative Results from Questionnaires in Portugal". In this paper, the author starts by describing the increasing impact that calls for "objectivity", as "a significant impulse of quantification", have had on political decision-making processes—especially in the US—since the 1930s. Then, he argues that scientific methods have been used increasingly in decision-making about technological innovations in order to create a "culture of objectivity", characterized as a "useful instrument to guide and improve" subjective decisions. However, based on a broad quantitative empirical study, Boavida shows that hitherto the expectation of "rational" guidance within those decision-making processes is strongly limited. Basically, these decision-making processes cannot, according to Boavida, be generalized. Both the type of technology and the type of innovation path usually define the social, political, and cultural contexts of decision-making processes. In order to understand these processes in a more generalized way, he claims it is necessary to broaden the decision-making model to have it embrace not just rational-analytical considerations but political-behavioral and emotional-intuitive issues too.

Part II: Extra-Disciplinary Transgression of Boundaries

Concerning the extra-disciplinary transgression of boundaries, Farah Purwaningrum, in her contribution "Shifting Practices of Academia as an Entrepreneurial Organization in Indonesia: The Case of ATMI Polytechnic Cikarang", describes knowledge production in vocational training systems. Based on her own empirical observations, she shows how the orientation on production processes is slowly changing the quality of academic teaching and—therefore—the quality of vocational training within the polytechnic organization. Although the vocational system of polytechnic structure has been strongly connected to manufacturing processes from the very beginning, recent trends like the liberalization of higher education in Indonesia exert enormous pressure on the survival of academia, which has been forced to shift its practices towards the production system. According to Purwaningrum, these shifts are the reason that tacit knowledge—which once formed a significant part of the production process—is losing relevance in teaching syllabuses and guidelines. From her perspective, this loss has negative effects on a student's level of qualification and, consequently, also negative effects on the quality of products in Indonesia's markets.

The next contribution in this part, Judith Igelsböck's "Designing 'Integration Machines': Computer Simulation and Modelling in Transdisciplinary Sustainability Research in Austria", is focused on a transdisciplinary Austrian research program on sustainability. This program calls for a heterogeneous production of knowledge on sustainability (where nonscientific actors are also invited to participate) through the use of computer modeling and simulations, or "integration machines", to use a term of the author. Based on an empirical analysis of the discursive contextualization and justification of transdisciplinary sustainability research provided by different research projects inside the program, the author exposes how traditional responsibilities, roles, and identities of scientific and social actors are rethought and redescribed. However, the analysis also demonstrates that integration machines do not just integrate, but also exclude some insights which never become part of heterogeneous knowledge production processes—and, in that sense, they can also be characterized as "antipolitics machines". The future-oriented perspective, trust in quantified knowledge, and the priority given to simulations and models are some of the boundaries that remain non-transgressable.

The starting point of the analysis in Genco Guralp's contribution, "Cosmology and Weberian Science", refers to Max Weber's famous "disenchantment" thesis, which he developed at the beginning of the twentieth century. When explaining the logic and the success of modern science, Weber strongly enhances the unreachable distance between science as an infinite methodological activity and issues of "meaning", i.e., issues about which science is not intended to provide irrevocable answers. In the context of cosmology and its latest developments, Guralp shows that this thesis is losing relevance. Although scientific evidence in cosmology is relatively controversial, the field, on the basis of an increasingly multidisciplinary constitution, has arguably broken with its "speculative" past and become a "precision science". Based on observations of different research groups in this field, Guralp focuses on the narratives of some outstanding cosmologists about issues such as the nature of "nothingness" or the existence of God. In their narratives, they are also asserting the "intellectual bankruptcy" of theology and—partly— philosophy concerning such issues. Assuming that physics recovers its "final theory" character, as claimed by some of these cosmologists, Guralp argues that the Weberian notion of the "infinite progress" of science is no longer plausible.

In the paper entitled "Cosmology and Theology: Some Mistakes in the Cosmological Case Against God", Fabrice Pataut, originally motivated by Genco Guralp's paper, claims that the theological and philosophical conclusions made on the basis of cosmological knowledge and theories are in principle illegitimate. He offers a critical analysis of the cosmology-based antitheistic arguments of Lawrence Krauss, Stephen Hawking, and Leonard Susskind, cosmologists who argue that we can do away with God in explanations of the origin and existence of the universe. Pataut, in contrast, argues that no inferences as to the best cosmological explanation can be used to justify the dispensability of the theological role that God plays in explaining the origin and existence of the universe because the two theories are incommensurable in terms of the type of explanation—naturalistic versus non-naturalistic—that they offer, and they do not even share theoretical terms. Thus, the author concludes that the extra-disciplinary transgression allegedly carried out by cosmologists is based on a misrepresentation of what cosmology can really reject as a scientific explanation.

Part III: Radical Transgression of Boundaries

This last part begins with Jaume Navarro's "Boundaries Between Territories of Knowledge. Colonization or Independence" paper, in which the author offers a deep characterization of Joseph John Thomson's (1856–1940) pluralistic stance on scientific knowledge during his early career. Thomson, who discovered the electron in 1897, was very critical of the tendency of his age to divide science into specialized disciplines, which he considered to be based on arbitrary and theoretically counterproductive boundaries. In contrast, as portrayed in the paper, he defended pluralism in science at different levels: Epistemological, ontological, and topical. He unsuccessfully tried to create a physical sciences department at his university (Cambridge) involving both physics and chemistry, tolerated a plural understanding of matter as both a continuous and discrete entity, and showed interest in psychic research, motivated by the possibility of extending physics to the study of the mind. In short, he went beyond advocating inter- or transdisciplinarity—based, in the end, on a sharp separation of disciplines—and claimed strong pluralism for science, always under the guidance of physics.

The next chapter, "Naturalism and Scientific Hierarchy: An Attempt at Strict Naturalist Normativity", by Pedro Sáez Williams, aims to overcome the assumed incompatibility between a naturalistic account of scientific knowledge and scientific normativity. In order to do that, the paper starts by criticizing the positivist attempts to characterize and justify science's cognitive authority on grounds of a non-contingent, or transcendental access to "facts"—or what the author terms as "the source". Interdisciplinary boundaries—which the author characterizes as "artificial barriers"—as well as hierarchical relationships between science and society are claimed to be based on that alleged transcendental or non-contextual objective access to reality. However, Sáez Williams claims that a normative framework for science can still be proposed on a strict naturalistic characterization of scientific knowledge. Based on the idea of "embodied cognition" from the cognitive sciences, knowledge is characterized as a relative—i.e., contingent to anatomy—but non-arbitrary process, constrained by embodiment. This naturalistic normativity is claimed here to have the potential to break down presumed hierarchies, or boundaries, in the sense that cognitive authority over others depends on the ability to transform the interlocutor's beliefs, or "limit of certainty", on non-hierarchical communicative grounds. The main philosophical mission here is to understand what kind of social organization would serve as "the most un-intermediated, unbounded and efficient form of communication".

Last but not least, Steve Fuller's "Prolegomena to a Genealogy of the Transgressive Mindset" exposes humanity's constitutive tendency toward transgression through history. More specifically, the paper analyzes the theological and philosophical roots of the transgressive mindset, which is characterized as being "about establishing a continuity of being between the human and the divine". On the theological side, the author focuses on Prometheus, Faust, and Simon Magus as exponents of humanity's aspiration to reach divinity—i.e., to overcome their natural limitations. In that sense, these theological myths are presented here as expressing an attempt to overcome absolute ontological boundaries, namely to understand differences between being human and being divine in terms of degree rather than in terms of kind. Concerning philosophy, the author focuses on Scotus' "univocal predication", in virtue of which "being" had for him the same meaning irrespective of what was said about it. This "gradualism", expressed by Scotus in terms of an explicit distinction between the hitherto undistinguished logical possibility and empirical probability, contributed to weakening the self-limiting dynamics of reason (i.e., "taboo cognition"), which tended to see itself as powerless to penetrate reality at certain levels. In the opinion of Fuller, Scotus, who "opened up humanity's epistemic horizons", prepared the ground for the seventeenth century Scientific Revolution, the main exponent of humanity's aspiration to know and control the world. In that sense, scientific transgression can be arguably interpreted as being part of the same theologico-philosophical transgressive impulse, meaning that a transgressor's boundaries are themselves also gradual and blurred.

Thus, the three parts present a broad variety of ways of reflecting on the boundaries of scientific disciplines in their social contexts. If we take into account the historical development of scientific disciplines, this observation should not be surprising. The boundaries between scientific disciplines have always been blurred, at least since the methodological constitution of sciences in the modern era (Daston and Lunbeck 2011). The contributions collected in this volume are aimed to specifically highlight the constitutive and contingent character of scientific boundaries. Intra-scientific as well as inter-scientific boundaries are versatile. Individual and institutional efforts at drawing sharp intra-scientific boundaries have been constant through the history of science. Moreover, even if science is still pretty much understood in our societies as a methodologically structured enterprise committed to truth and progress, the history of science and scientific breakthroughs cannot be understood without taking into account less structured dimensions. These are methodological failures, accidents, temporal and spatial shifts, individual scientists' characters and efforts, or even simple luck (Daston and Lunbeck 2011). The description of scientific action as a social process has relatively eroded the idea of "objective" science.

As seen throughout this introduction, though, the issue of extra-scientific boundaries has probably attracted more attention than either the intra- or inter-scientific ones both in the scientific and in the public arena. The responsibility of solving societal problems is increasingly being delegated to the sciences, and so scientific boundaries are becoming contingent as well as socially legitimate in virtue of that problem-solving function. Thus, as already seen, the contingency of scientific boundaries is interrelated with the "contingencies of particular historical situations" (Barnes et al. 1996, p. 140): They demand deep reflection on the demarcation between scientific boundaries and societal needs, interrogating us about the role played by scientific knowledge in our societies currently and through history.

Disciplinary demarcations are thus arguably conventional, relative to their social and historical contexts (Kuhn 1962; Daston and Galison 2007). They respond to specific, situated goals and strategies. Barnes and colleagues, for instance, argue that boundaries are mainly created to satisfy and maintain some specific interests of scientists, "concerned to protect and promote their cognitive authority, intellectual hegemony, [and] professional integrity" (Barnes et al. 1996, p. 168). As far as science becomes the addressee for resolving societal problems, the issues of methodological and ethical integrity become crucial, constituting a defining characteristic of an increasingly contested scientific activity (Latour 2013).

Thus, the role and the function of science in society are under steady observation. We hope that the current publication contributes to promoting the debate and our understanding of the relevance and meaning of scientific boundaries.

Acknowledgments Hannot Rodríguez's contribution is based on research supported by a personal fellowship from the Karlsruhe Institute of Technology (KIT) and conducted at the Institute of Technology Assessment and Systems Analysis (ITAS). It is also based on research supported by the Spanish Ministry of Economy and Competitiveness under grants FFI2011-24414 and FFI2012-33550, the Basque Government's Department of Education, Language Policy and Culture under grant IT644-13, and the University of the Basque Country UPV/EHU under grant EHUA15/13.

References

Allenby BR, Sarewitz D (2011) The techno-human condition. MIT Press, Cambridge

Barnes B, Bloor D, Henry J (1996) Scientific knowledge: a sociological analysis. Athlone Press, London

Bauer M (ed) (1995) Resistance to new technology: nuclear power, information technology and biotechnology. Cambridge University Press, Cambridge

Bechmann G, Decker M, Fiedeler U, Krings B-J (2007) Technology assessment in a complex world. Int J Foresight Innovation Policy 3(1):6–27

Berlin I (1990) The crooked timber of humanity: chapters in the history of ideas. Edited by Henry Hardy. John Murray, London

Black J (2013) War and technology. Indiana University Press, Bloomington

Boudet H, Clarke C, Bugden D, Maibach E, Roser-Renouf C, Leiserowitz A (2014) "Fracking" controversy and communication: using national survey data to understand public perceptions of hydraulic fracturing. Energy Policy 65:57–67

Bruce A, Lyall C, Tait J, Williams R (2004) Interdisciplinary integration in Europe: the case of the fifth framework programme. Futures 36(4):457–470

Busquin P (2003) Foreword: building a knowledge-based Europe—our common challenge. In: European Commission, Research DG. The overall socio-economic dimension of community research in the 5th European framework programme—A synthesis report on the integration of the socio-economic related research activities of the European Community (1998–2002) (EUR 20577). Office for Official Publications of the European Communities, Luxembourg, p. 6

Callon M, Lascoumes P, Barthe Y (2009) Acting in an uncertain world: an essay on technical democracy (translated by G. Burchell). MIT Press, Cambridge/London

Cheek DW (2005) Hiroshima and Nagasaki. In: Mitcham C (ed) Encyclopedia of science, technology, and ethics. Macmillan Reference, USA, Detroit, pp 921–924

Cozzens S, Wetmore J (eds) (2010) Nanotechnology and the challenges of equity, equality and development, Springer, New York

Crystal S, Siegel MJ (2009) Population aging and health care policy in cross-national perspective. In: Uhlenberg P (ed) International handbook of population aging. Springer, Dordrecht/London, pp 607–630

Daston L, Galison P (2007) Objectivity. Zone Books, Brooklyn, NY

Daston L, Lunbeck E (eds) (2011) Histories of scientific observation. University of Chicago Press, Chicago/London

Dawkins R (2008) The god delusion. Houghton Mifflin Co., Boston

Dennett DC (1995) Darwin's dangerous idea: evolution and the meanings of life. Simon & Schuster, New York

Dennett DC (2006) Breaking the spell: religion as a natural phenomenon. Viking, New York

Douglas H (2000) Inductive risk and values in science. Philos Sci 67(4):559–579

EC (2001) European governance: a white paper. Brussels, 25.7.2001, COM(2001), 428 final

EC (2007) Work programme 2007—capacities, Part 5: science in society (C(2007)563 of 26.02.2007). The Seventh Framework Programme, Brussels

EC (2011) Horizon 2020—The framework programme for research and innovation. Brussels, 30.11.2011, COM(2011) 808 final

EC (2014a) The European Union explained: research and innovation. Publications Office of the European Union, Luxembourg

EC (2014b) Science with and for society (Revised). Horizon 2020, work programme 2014–2015, European Commission Decision C (2014)4995 of 22 July 2014

Felt U (2013) Keeping technologies out: sociotechnical imaginaries and the formation of national and technopolitical identity. Department of Social Studies of Science, University of Vienna, Vienna Preprint

Fisher E (2007) Ethnographic invention: probing the capacity of laboratory decisions. NanoEthics 1(2):155–165

Fisher E, Mahajan RL (2006) Contradictory intent? US federal legislation on integrating societal concerns into nanotechnology research and development. Science and Public Policy 33(1):5–16

Flipse SM, van der Sanden MC, Osseweijer P (2013) Midstream modulation in biotechnology industry: redefining what is "part of the job" of researchers in industry. Sci Eng Ethics 19(3):1141–1164

Fukuyama F (2002) Our posthuman future: consequences of the biotechnology revolution. Farrar, Straus and Giroux, New York

Fuller S (2011) Humanity 2.0: What it means to be human past, present and future. Palgrave Macmillan, Basingstoke

Funtowicz SO, Ravetz JR (1993) Science for the post-normal age. Futures 25(7):739–755

Gaskell G (2008) Lessons from the bio-decade: a social scientific perspective. In: David K, Thompson PB (eds) What can nanotechnology learn from biotechnology? Social and ethical lessons for nanoscience from the debate over agrifood biotechnology and GMOs. Academic Press, Amsterdam, pp 237–259

Gibbons M, Limoges C, Nowotny H, Schwartzman S, Scott P, Trow M (1994) The New Production of Knowledge. The Dynamics of Science and Research in Contemporary Societies, Sage, London

Giddens A (2009) The politics of climate change. Polity Press, Cambridge/Malden

Grunwald A, Schmidt J (2005) Method(olog)ische Fragen der Inter- und Transdisziplinarität. Technikfolgenabschätzung—Theorie und Praxis (TATuP) 2:5–11

Guston DH, Sarewitz D (2002) Real-time technology assessment. Technol Soc 24(1–2): 93–109

Guston DH, Fisher E, Grunwald A, Owen R, Swierstra T, van der Burg S (2014) Responsible innovation: motivations for a new journal. J Responsible Innovation 1(1): 1–8

Jasanoff S (1990) The fifth branch: science advisers as policymakers. Harvard University Press, Cambridge

Jasanoff S (2004) States of knowledge: the co-production of science and social order. Routledge, London

Jasanoff S, Kim SH (2009) Containing the atom: sociotechnical imaginaries and nuclear power in the United States and South Korea. Minerva 47(2):119–146

Kondo N, Monta M, Noguchi N (eds) (2006) Agricultural robots. Kyoto University Press, Kyoto, Mechanisms and Practice

Krauss LM (2012) A universe from nothing: why there is something rather than nothing. Free Press, New York

Krings B-J (ed) (2011) Brain drain or brain gain? Changes of work in knowledge-based societies. Sigma, Berlin

Kuhn TS (1962) The structure of scientific revolutions. University of Chicago Press, Chicago

Latour B (2013) An inquiry into modes of existence. An anthropology of the moderns. Harvard University Press, London

Longino HE (1990) Biological effects of low level radiation: values, dose-response models, risk estimates. Synthese 81(3):391–404

Macnaghten P (2008) From bio to nano: learning the lessons, interrogating the comparisons. In: David K, Thompson PB (eds) What can nanotechnology learn from biotechnology? Social and ethical lessons for nanoscience from the debate over agrifood biotechnology and GMOs. Academic Press, Amsterdam, pp 107–123

Marklund G, Vonortas NS, Wessner CW (eds) (2009) the innovation imperative: national innovation strategies in the global economy. Edward Elgar, Cheltenham

Meadows DH, Meadows DL, Randers J, Behrens WW (1972) The limits to growth. Universe Books, New York

Millennium Ecosystem Development (ed) (2005) Ecosystems and human well-being: synthesis report. World Resources Institute, Island Press, Washington

Nelkin D (1994) Science controversies: the dynamics of public disputes in the United States. In: Jasanoff S, Markle GE, Petersen JC, Pinch T (eds) Handbook of science and technology studies. Sage, Thousand Oaks, pp 444–456

Nowotny H (2005) Unersättliche Neugier: Innovation in einer fragilen Zukunft. Kulturverlag Kadmos, Berlin

NSF (2011) Empowering the nation through discovery and innovation—NSF strategic plan for fiscal years (FY) 2011–2016. (NSF 11-047) Arlington, VA

NSTC (2014) National nanotechnology initiative—strategic plan. Executive Office of the President of the United States, Washington

O'Hear A (ed) (1996) Verstehen and humane understanding. Cambridge University Press, Cambridge

Parayil G, D'Costa AP (2009) The new Asian innovation dynamics: China and India in perspective. Palgrave Macmillan, Basingstoke/New York

Perrow C (1984) Normal accidents: living with high-risk technologies. Basic Books, New York

Pillemer K (2011) Environmental sustainability in an aging society: a research agenda. J Aging Health 23(3):433–453

Rabinow P (1996) Making PCR: a story of biotechnology. University of Chicago Press, Chicago

Rescher N (1999) Razón y valores en la era científico-tecnológica (edited by W.J. González) [Reason and values in the scientific-technological age]. Paidós/ICE-UAB, Barcelona, Barcelona

Rhoten D (2004) Interdisciplinary research: trend or transition. Items Issues 5(1–2):6–11

Ridley M (2010) The rational optimist: how prosperity evolves. Harper, New York

Rodríguez H (2016) From objective to constituted risk: an alternative approach to safety in strategic technological innovation in the European Union. J Risk Res 19(1):42–55

Rodríguez H, Fisher E, Schuurbiers D (2013) Integrating science and society in European Framework Programmes: trends in project-level solicitations. Res Policy 42(5):1126–1137

Saffiotti A, Lima P (eds) (2008) Two "hot issues" on cooperative robotics: network robotic systems, and formal models and methods for cooperation. White paper from the EURON Special Interest Group on Cooperative Robotics

Sandel MJ (2007) The case against perfection. Harvard University Press, Cambridge

Sarewitz D (1996) Frontiers of illusion: science, technology and the politics of progress. Temple University Press, Philadelphia

Schot J, Rip A (1997) The past and future of constructive technology assessment. Technol Forecast Soc Chang 54(2–3):251–268

Schuurbiers D (2011) What happens in the lab: applying midstream modulation to enhance critical reflection in the laboratory. Sci Eng Ethics 17(4):769–788

Stirling A (2008) "Opening up" and "closing down": power, participation and pluralism in the social appraisal of technology. Sci Technol Human Values 33(2):262–294

Tokamanis C (2011) Nanotechnology becomes a socio-political project (Interview). In: Bonazzi M (ed) Successful European nanotechnology research. Outstanding science and technology to match the needs of future society. Publications Office of the EU, Luxembourg, pp 9–12 van Rijnsoever FJ, Hessels LK (2011) Factors associated with disciplinary and interdisciplinary research collaboration. Res Policy 40(3):463–472

von Hippel E (2005) Democratizing innovation. MIT Press, Cambridge

Watts-Roy DM, Williamson JB (2009) Public pension programs—social security. In: Uhlenberg P (ed) International handbook of population aging. Springer, Dordrecht/London, pp 427–428

Weinert F (2009) Copernicus, Darwin & Freud: revolutions in the history and philosophy of science. Wiley-Blackwell, Chichester/Malden

Wright SD, Lund DA (2000) Gray and green? Stewardship and sustainability in an aging society. J Aging Stud 14(3):229–249

Part I
Disciplinary Transgression
of Boundaries

Science and Experiment

Hans-Jörg Rheinberger

Abstract

The chapter analyses procedural aspects of experimentation that are commonly considered as "subjective" or "contingent" and ignored because they seem to be inconsistent with the claim to authority of scientific action. Based on a description of experimentation as a type of open exploration of the world, the role of the principles of uncertainty and serendipity in producing epistemic things is discussed. The chapter shows how the unexpected emergence of something new in the history of science mostly takes place between the poles of technical and epistemic chance. It locates the aesthetic moment of experimentation precisely in the combination of complexity and manageability of limited, but nevertheless multi-layered systems. For instance, an experimental system would not—as is often stated—be perceived as beautiful or aesthetic because it is "simple", but because it hovers "on a borderline". Its structure is repetitive but also holds surprises.

Keywords

Experimentation · Experimental systems · Serendipity · Molecular biology · Epistemic things

The present text was translated by Michael Wilson. It is a slightly modified version of a lecture given under the title "Wissenschaft und Experiment" and published in: Anne von der Heiden and Nina Zschocke (eds.) (2012) *Autorität des Wissens. Kunst- und Wissenschaftsgeschichte im Dialog*, Zurich, pp. 123–133. See also Rheinberger (2006a).

H.-J. Rheinberger (✉)
Berlin, Germany
e-mail: rheinbg@mpiwg-berlin.mpg.de

© Springer Fachmedien Wiesbaden GmbH 2016
B.-J. Krings et al. (eds.), *Scientific Knowledge and the Transgression of Boundaries,* Technikzukünfte, Wissenschaft und Gesellschaft / Futures of Technology, Science and Society, DOI 10.1007/978-3-658-14449-4_2

This paper deals with aspects of experimenting in modern science that, as a rule, tend to be *attributed* to subjective and contingent factors and thus are largely excluded from the sociological and, in particular, the philosophical study of experimentation.[1]

1   Research as Searching

Without a doubt, research is the core activity of modern science, and one could add that *experimenting* makes up the core activity of modern research. Research can be characterized as a searching that takes place at the border between knowledge and our lack of knowledge (Rheinberger 2005). The fundamental problem consists in the fact that one does not exactly know what one does not know. This is a brief but precise description of the core of scientific research. The goal is ultimately to acquire *new* knowledge, and yet by definition whatever is truly new cannot be anticipated. There are therefore limits to our capacity to bring it about. Whatever is truly new must simply happen; it must take place. In an experiment, a researcher creates an empirical structure, or environment, that enables him to act in this state of not knowing about the unknown. Yet, an experimental setup does embody a large amount of knowledge that at a certain point in time is considered to be verified and that, as a rule, takes the form of instruments, devices, and apparatuses. The latter are in fact often set in motion solely to confirm that they are in good working order; the calibrating and testing of apparatus probably even takes up the largest part of a scientific experimenter's time. The machines that are employed are supposed to do their work smoothly and silently. The actual goal of experimenting consists in getting the phenomena that are being examined to speak up. In my terminology, I refer to these phenomena as the epistemic things (Rheinberger 2001). The exploratory experiment has to be set up in such a manner that the unexpected can take place. Claude Bernard once noted in one of his notebooks (Bernard 1965, p. 145): "It has been claimed that I would find something that I hadn't been looking for while Helmholtz only finds what he looks for. That is true, but such exclusiveness is bad in either direction". This statement by the great French physiologist of the nineteenth century—independent of his indirect swipe ("it has been claimed") at his German colleague Hermann von Helmholtz—hits the decisive point precisely. An experiment that yields findings is designed to reveal

---

[1]Some of the exceptions are Pierce (1955) *Philosophical Writings* (see here in particular Chap. 11: Abduction and Induction) and Merton (1949) *Social Theory and Social Structure*.

something that the experimenter does not yet have a precise conception of. On the other hand, the experimenter cannot be surprised by something new if he does not have at least a vague conception of it. The nature of the experimental spirit must therefore be complementary to the experimental structure. Researcher and object form a tight relationship: The better he knows "his thing", the more subtly it shows itself to him. The experiment is, so to speak, a search engine, but one with an unusual structure. It namely produces things of which the experimenter can always only subsequently say that he would have had to have looked for them. Bernard is thus correct when he categorically concludes, "Knowledge is always something a posteriori" (Bernard 1954, p. 21).

The history of science is full of stories about such surprising findings. Ludwik Fleck once even referred to them as the Columbus effect. Expressed rather informally: You look for India, and you find America (Fleck 1979, p. 69). This frequently mentioned example reveals something else. When the new item makes its first appearance, it is usually not recognized as new, but becomes something new in a recursive process. As far as we know, Columbus died believing he had found new regions in the orient—to which he had sought an alternative route—and not an unknown continent. Examples of chance discoveries in the more recent history of science that have been cited over and over again are the discovery of gamma rays by Wilhelm Conrad Röntgen and of penicillin by Alexander Fleming.

2 Serendipity

In the course of the last few decades, the term "serendipity" (or "the serendipity principle") has been used commonly to refer to this circumstance (see Roberts 1989; Merton and Barber 2003). An English eighteenth century author, Horace Walpole, is credited with having coined the term. In a letter to his friend Horace Mann in 1754, he talked at length about a Persian fairy tale with the English title *The Three Princes of Serendip*. By a combination of chance and acumen, the three princes in the fairy tale were constantly discovering things they had not been looking for (Roberts 1989, p. ix). The principle of discovery described here is thus not pure coincidence. It describes an event that takes place without one moving directly toward it but that on the other hand does not simply cross one's path. We could perhaps speak of a canalized or facilitated chance occurrence. Two qualities are needed in order to bring about this occurrence as well as to grasp it: A familiarity in dealing with the materials that only comes with experience, and a type of attentiveness that contains a sharp sense for distinguishing nuances. This is thus acumen that may not be too rigidly focused but that can, as it were, be held in suspension. Roberts (1989, p. x) made the effort to distinguish between serendipity and pseudoserendipity. According to him, the category of pseudoserendipity is for discoveries of new and unexpected procedures for achieving a certain goal that the researcher firmly has in mind. The category of genuine serendipity, in contrast, is reserved for those discoveries that concern things that one is precisely not looking for. One could accordingly distinguish something like technical coincidences from epistemic ones. Presumably, however, every concrete historical example can be located somewhere in the wide gray area between these two extremes.

Two examples from the history of science may, to begin with, illustrate these extremes. The research conducted by Claude Bernard, mentioned above, about the *breakdown* of sugar in the body led at a critical point in his experiments in the 1840s, after several years of experimental work, to the totally unexpected observation that an animal's body is also capable of *synthesizing* sugar (Holmes 1974). As a consequence of this reversal—an example of epistemic serendipity—fundamental changes took place in our thinking about animal metabolism. As the German zoologist Alfred Kühn was busy developing his experimental system of genetics oriented on developmental physiology and based on the Mediterranean flour moth (*Ephestia kuehniella*), his massive breeding program produced a red-eyed mutant of the moth, detected because of the attentiveness of one of his technical assistants. This technical serendipity became the starting point of all the further studies of Kühn's research group on gene action chains affecting chains of metabolic substrates, and it formed the basis of biochemical genetics that subsequently developed (see Rheinberger 2006b). This example clearly demonstrates, however, the dubiousness of the distinction. In the last case, we can say with some justification that it was precisely the new mutant that made it possible to develop the distinction between gene networks and substrate chains, which in turn had a massive epistemic impact.

As a rule, such events become the topic of scientific legends and anecdotes, and accordingly great care must be exercised in the history of science. This is the reason we rarely find them formulated explicitly in original scientific publications; they appear rather in autobiographic recollections, where as a rule they are reported trenchantly from the perspective of later successes. The serendipitous good fortune of a scientist is itself actually something that can only be appreciated after the fact. It is always on the—potentially dangerous—brink of minimizing the role of the person who happens to experience it because the good fortune is nothing that one has earned. For this reason, consequently, one does not speak about it until after achieving subsequent results, i.e., after having taken the next steps. A good example of this is Eduard Buchner, one of the founders of in vitro biochemistry around the turn of the twentieth century. He begins his legendary paper on

"Alcoholic Fermentation without Yeast Cells" ("Alkoholische Gärung ohne Hefezellen") from 1897 by writing succinctly: "Separation of fermentation from living yeast cells has not been achieved previously. In the following, I describe a procedure that solves this task" (Buchner 1897). Buchner did not begin to speak about the circumstances of his discovery until from a safe distance, and even this was triggered by claims that an assistant had made to the discovery. A recent biography describes in detail that Buchner—impressed by the research done by his brother Hans Buchner—was actually searching for an immunizing substance to be obtained from microorganisms (Ukrow 2004). In order not to modify the presumed substance *chemically*, he turned to the *mechanical* destruction of the cells. To stabilize the resulting unstable cell sap, he first added glycerin and then tentatively sugar. It was only the foaming of the resulting liquid that directed his attention to the fermentation process that was to determine his scientific work for the next decade.

A serendipitous event often has the character of something that—at least from the perspective of the experiment just conducted—has gone wrong or, in other words, that an assumption was not confirmed. In a nutshell, serendipity means to turn this "misfortune" into a productive event and thus, depending on the circumstances, turning the entire experimental undertaking in a new and unexpected direction. A special form of this occurs when what is intended to be a control experiment is turned into the actual one. Employing a new instrument, the ultracentrifuge, the Belgian physician Albert Claude succeeded at the Rockefeller Institute in New York, around the middle of the 1930s, in concentrating the active, carcinogenic principle from sarcoma cells of chickens several thousand times. Yet when he centrifuged sap from healthy chicken tissue to check his results, he determined to his surprise that the inactive sediment from the control experiment did not differ in its composition from the active specimen. This led him to become interested in the source of this sediment, instead of in the carcinogenic principle (Löwy 1990). His subsequent experiments turned into one of the starting points of experimental in vitro cytology in the second third of the twentieth century. In my book on the history of research into protein synthesis, I have described in detail another such experiment that was conducted to confirm results and that marks a biochemical turning point in molecular genetics. The use of a radioactive amino acid in parallel to that of a radioactive nucleotide led Paul Zamecnik, at the Massachusetts General Hospital in Boston in the mid-1950s, to characterize a small, soluble nucleic acid that initially had appeared to represent an ineradicable contamination of the experimental system. As it became apparent that an amino acid could be transferred to this nucleic acid, a carrier of the transfer of genetic information began to take on shape at the molecular level, a perspective that had not at all been the intention of the original experiment (Rheinberger 2001).

Even these few examples demonstrate that such turns in research can take completely different forms. Technical accidents can be the reason that previously neglected phenomena come to light. In this case, a mishap becomes a productive factor. A control experiment can be transformed into a research experiment. In this case, an unchallenged assumption—it is in the nature of controls that they embody the respective state of knowledge in compact form—becomes a problem. The procedures that are employed can result in effects other than those originally intended. In this case, a resource (e.g., instrument or act) produces an unanticipated excess effect. Components that are considered contaminants of a system may prove to be integral parts that cannot be removed and that are transformed from a disturbance into an object of study. There are cases of surprising incidental results, whose significance depends on their even being noticed. Although to my knowledge no typology of unanticipated and unanticipatable turns in experiments has yet been written, we can follow Root-Bernstein (1989, p. 365) in asserting that in science "without experiments with serendipitous results soon all theorizing would come to a halt". And Root-Bernstein (1989, p. 376) summarizes: "Science *is* change", in fact in the sense of "actual, effective surprise". In this, he refers to the conviction of the English philosopher of science Stephen Toulmin that what is new in science is just as often unexpected and occurs in an unintended manner as is the case in nature (Toulmin 1961).

Experimental systems can be considered as structures that make it possible for such turns to take place during the process in which we expand our knowledge. In other words, they are structures that permit us to assimilate coincidences in a productive manner and that perhaps are a prerequisite for creating a form of coincidence that can be assimilated in this fashion. This arrangement thus becomes the focal point of all science in the making or, to use the phrase of Latour (1987), "science in action". As Bernard (1965, p. 135) wrote in his red notebook, "Where you do not know further, there you have to find". He noted at a different spot: "You can probably say that no one has yet made a discovery by looking for it directly" (Bernard 1965, p. 149). The experiment is the form in which modern science has imposed rules—which themselves are subject to historical change—on this indirect, searching manner of finding. Ultimately, something imponderable is left over. This is inextricably associated with the diversity of elements that make up an experimental set-up. These elements are both material in nature as well as social, cultural, and epistemic. It is impossible to specify an ideal type or an ideal mixture. As Bernard wrote in one of his notes:

Everyone goes their own way. Some prepare themselves long beforehand and march along a predetermined path. I myself reached the field of science via detours, and I rid myself of the rules by finding a place between the disciplines, something others may not have dared to do. Yet I believe this has not hurt in the field of physiology since it has led me to have new insights (Bernard 1965, pp. 128–129).

Individual disciplines are always at different stages of formation, and accordingly particular strategies of research prove to be more or less successful. And every experimental system is ultimately concrete, homemade, and in an elementary sense a product of its time and means.

Yet there is also a flipside to this, i.e., those configurations in which an experimental clue was overlooked and which in hindsight appeared to the participants to be a missed chance. The history of science is also rich in examples of missed chances. In retrospect, for example, a half dozen researchers and research groups could have described messenger RNA, which was so decisive for the establishment of molecular genetics. Often, events depend on which system one chooses. If the genetic code had not been solved in a bacterial system of in vitro protein synthesis, those researchers who worked with mutants of the tobacco mosaic virus might have been successful. The cases that are particularly aggravating for the competitors are those in which they almost simultaneously and independently reach a result that subsequently proves to be pathbreaking. This is particularly annoying where priority is determined by tiny details in the perception of its significance and by changes in a journal's publication schedule. As with other creative activities, research can also be a risky game, and a detail that appears relatively minimal can be decisive for careers.

3 In the Right Place at the Right Time

This leads us to an aspect of research that is somewhat different, situated at a different level, but associated with the preceding comments. Scientists often refer to it by using the expression of having the good fortune of having been in the right place at the right time. This is a reference to two essential features of experimental systems. First, as a rule, such systems are locally adapted, and in this local adaption they may be unique, i.e., not transplantable at one's will. Secondly, they possess something like an internal dynamic, which can always lead to critical inflection points at which a relatively minor input from outside can trigger large movements. Mahlon Hoagland joined Paul Zamecnik's group in Boston in the 1950s after working with Fritz Lipmann, with whom he had learned the tools of the trade in phosphorylation experimentation by working on the nucleotide adenosine triphosphate (ATP). The application to Zamecnik's cell-free system for incorporating amino acids in proteins led in less than a year to the result—which was pioneering for the further development of the system—that amino acids have to be activated by ATP before they can be utilized in protein synthesis (Hoagland 1990). Yet this outside input does not always have to be technical in nature. In his autobiography *La statue intérieure*, François Jacob described a particular conjuncture of being in the right place at the right time. At the beginning of the 1950s, he joined the group led by André Lwoff and Elie Wollman at the Institut Pasteur at a time when Lwoff's longstanding interest in lysogenic phages was developing into an entry point to experimental bacterial genetics, which Jacob then successfully made his own cause.

This type of a turn in research is closely tied to a scientist's biography and is naturally also linked to decisions at the level of institutions. For the individual researcher, it is his personal increment, i.e., the stamp he puts on an experimental system that as a rule is already there. From the perspective of the development of the *system*, the researcher's entrance itself becomes an element in the play of its contingencies: That he hit on it, and that it hit him of all people. This also shows something of the texture of the experimental sciences, which as a rule are made up of many smaller, patchwork-style units that unite to form an experimental landscape. Between the regions of this landscape there may be a limited, selective exchange of persons, objects, instruments, and procedures. It is often precisely these cross-links—the elements that a researcher contributes to an existing system—that catalyze processes that may have been inherent in a certain experimental project but may not be recognized by those immediately involved. Close familiarity with an experimental system is on the one hand a necessary condition for its productive manipulation, but on the other hand it can also constrict the view of those involved. In this connection, the semi-controlled contingency of the exchange of personnel in a scientific community plays a decisive role in releasing the potential that is inherent in a particular research process.

4 Virtuosity

In conclusion, I would like to present a short look at the subjective side of craftsmanship in research. This is tied to the performance of scientific experimental work and implies both aesthetic and kinaesthetic aspects. In his book on Emil Du Bois-Reymond, Dierig (2006) shows that Du Bois-Reymond combined an understanding of sound experimental work in the tradition of a craftsman with an aesthetics of functionality or of functional form that presaged the approaching industrial age. The subjective side also included the athletically trained, fit body of the experimenter geared toward the Greek ideal (see the exhibition catalogue edited by Dierig and Schnalke 2005). Du Bois-Reymond spent his entire life consummating this experimental activity that was directed at perfection in design and virtuosity in handling. This was the case both in the completion of his electrophysiological studies and in the demonstrations he presented in lectures to a large public. It can also be said that this self-realization in experimenting that was stylized to an ideal and this attitude of seeking self-satisfaction in virtuosity were not solely to the benefit of science. Du Bois-Reymond's manner of pursuing electrophysiology was revolutionary in the 1840s, but had outlived itself by the 1880s. Virtuosity is no guarantee for productive research.

On the contrary, limited accuracy could help, being an attenuated form of precision as it were. If we follow the self-portrayal of molecular biology for which Horace Judson's book *The Eighth Day of Creation* (1979) set a memorial, then Max Delbrück, a physicist who turned to phage research, spoke in this connection of a "principle of limited sloppiness" (Fischer 1988). If someone handles his system with too much virtuosity and narrows it too far, nothing will come of it at the end. However if one lets chance play a role in a controlled manner, then he has a chance of finding something new. Delbrück's statement was, if nothing else, an expression of the transdisciplinary culture of molecular biology in the middle of the twentieth century that was oriented toward the transfer of methods, exploration, and new epistemic things. This was an alternative scientific culture that established itself beyond the traditional disciplines and in self-confident disregard of traditional boundaries.

This new orientation did nothing to lessen the satisfaction provided by a skilled experiment. Judson relates as follows an adage handed down from Alfred Hershey, who succeeded in proving at the beginning of the 1950s that the active principle of T-phages was its DNA and not its proteins. Hershey was once asked by Alan Garen, a colleague, what he imagined to be a scientist's greatest happiness. His answer was, "To have an experiment that works, and to do it over and over again" (Judson 1979, p. 196). François Jacob, however, presents this condition—which has come to be known to molecular biologists as Hershey heaven—in a significantly more differentiated form, which reminds one less of Du Bois-Reymond and is more compatible with the Delbrück principle:

> Al Hershey, one of the most brilliant American specialists on bacteriophage, said that, for a biologist, happiness consists in working up a very complex experiment and then repeating it every day, modifying only one detail (Jacob 1988).

The point in Jacob's version is thus not the unchanging repetition of a functioning experiment, not virtuosity à la Du Bois-Reymond, but that the potential of a complicated system can only be sounded out by trying it out in all directions and that this sampling of a possibility space can trigger a feeling of satisfaction of the experimenter's wishes. The aesthetic aspect here is in the combination of the complexity and the operability of a system that—while limited—is still complex enough and in which surprises are possible, i.e., in the marginal connection between presumed success and the unguaranteed hope for unanticipated answers. Thus the system is not considered beautiful because it is simple—in the sense of the elegant, simple solutions often invoked in the scientific literature. It is because it moves on a boundary: It has a structure that can be repeated and does not dissolve, while it also offers unsolved issues. The kinaesthetic aspect is in experiencing the virtuosity of experimenting that does not strive to be glossy, but rather to experience the independent feeling of being able to work with the material in a sovereign manner. It takes a great effort to get this far, as Hershey himself knew. If one has overcome the difficulties and set up a functioning experimental system, then one basically does not want to leave it any more. The experimenter has arranged himself in it, knows his way around in it, but cannot look behind every corner of the set-up, and he wants to inspect the full radius of the mute knowledge that is enclosed within his interaction with the system—his epistemic complicity.

References

Bernard C (1954) Philosophie. Manuscrit inédit. Hatier-Boivin, Paris

Bernard C (1965) Cahier de notes 1850–1860. Gallimard, Paris

Buchner E (1897) Alkoholische Gärung ohne Hefezellen. Berichte der Deutschen Chemischen Gesellschaft 30:117–124

Dierig S (2006) Wissenschaft in den Maschinenstadt. Emil Du Bois-Reymond und seine Laboratorien in Berlin. Wallstein Verlag, Göttingen

Dierig S, Schnalke T (eds) (2005) Apoll im Labor. Bildung, Experiment, Mechanische Schönheit. Berliner Medizinhistorisches Museum, Berlin

Fischer EP (1988) Das Atom der Biologen. Max Delbrück und der Ursprung der Molekulargenetik. Piper, Munich

Fleck L (1979) Genesis and development of a scientific fact. University of Chicago Press, Chicago

Hoagland M (1990) Toward the habit of truth. A life in science. W.W. Norton & Company, New York

Holmes FL (1974) Claude Bernard and animal chemistry. Harvard University Press, Cambridge

Jacob F (1988) The statue within. An autobiography. Basic Books, New York

Judson HF (1979) The eighth day of creation. Makers of the revolution in biology. Simon & Schuster, New York

Latour B (1987) Science in action. How to follow scientists and engineers through society. Harvard University Press, Cambridge

Löwy I (1990) Variances in meaning in discovery accounts. The case of contemporary biology. Hist Stud Phys Biol Sci 21:87–121

Merton RK (1949) Social theory and social structure. Toward the codification of theory and research. Free Press, Glencoe

Merton RK, Barber E (2003) The travels and adventures of serendipity. A study in sociological semantics and the sociology of science. Princeton University Press, Princeton

Pierce CS (1955) Philosophical writings. Dover, New York

Rheinberger HJ (2001) Experimentalsysteme und epistemische Dinge. Eine Geschichte der Proteinsynthese im Reagenzglas. Wallstein, Göttingen

Rheinberger HJ (2005) Nichtverstehen und Forschen. In: Albrecht J, Huber J, Imesch K et al (eds) Kultur Nicht Verstehen. Springer, Zurich, pp 75–81

Rheinberger HJ (2006a) Glück in der Forschung. In: Breuninger R (ed) Bausteine zur Philosophie, vol 26. GlückHumboldt-Studienzentrum, Ulm, pp 95–109

Rheinberger HJ (2006b) Epistemologie des Konkreten. Studien zur Geschichte der modernen Biologie. Suhrkamp, Frankfurt am Main, pp 131–184

Roberts RM (1989) Serendipity. Accidental discoveries in science. Wiley, New York

Root-Bernstein RS (1989) Discovering. Inventing and solving problems at the frontiers of scientific knowledge. Harvard University Press, Cambridge

Toulmin S (1961) Foresight and understanding. An inquiry into the aims of science. Harper & Row, New York

Ukrow R (2004) Nobelpreisträger Eduard Buchner (1860–1917). Ein Leben für die Chemie der Gärungen und—fast vergessen—für die organische Chemie. Dissertation, TU Berlin

From "Imaging 2.0" to "Imaging 3.0"

On the Crises of Radiology and Its "Culture Shifts"

Kathrin Friedrich

Abstract

Two major technological turnovers led to a crisis in the disciplinary self-conception of clinical radiology within the last 20 years. In the late 1990s, the transformation of analogue visualization techniques and diagnostic workflows into all-digital environments challenged the radiological modes of processing and reading medical images. This so-called analogue-digital migration fundamentally transformed the diagnostic process into a fully digital one. The analysis of user modeling processes and of its implementation in everyday work routines demonstrates the interdisciplinary tensions between radiology and, in particular, software design on how to conceptualize a discipline and its novel digitally-based work practices. More recently, the use of imaging techniques and visualization technologies in other medical disciplines causes strategic and political campaigns such as Imaging 3.0 to claim radiology's very own professional position in changing clinical environments and in health care policy decisions. Radiology tries to demand back its leading role as a clinical discipline with long-lasting traditions of visual expertise by simultaneously opening to new political demands such as patient-centered health care. The analysis of tumor boards illustrates the multidisciplinary contexts of today's cancer treatment regimes and thereby points to the contested status of radiology as a discipline with a very specific visual expertise. I will explore the professional "crises" and proclaimed culture shifts of radiology in

K. Friedrich (✉)
Berlin, Germany
e-mail: kathrin.friedrich@hu-berlin.de

© Springer Fachmedien Wiesbaden GmbH 2016   35
B.-J. Krings et al. (eds.), *Scientific Knowledge and the Transgression of Boundaries,* Technikzukünfte, Wissenschaft und Gesellschaft / Futures of Technology, Science and Society, DOI 10.1007/978-3-658-14449-4_3

the course of the analogue-digital migration as well as the Imaging 3.0 campaign by focusing on the relations between conceptual developments on a more macroscopic level and their impact on everyday practices within radiological settings.

Keywords

Technological innovation · User-modeling · Image-guided intervention · Visual expertise · Clinical radiology

1 Introduction

The future of radiology is bright; the future for radiologists is far less certain (Muroff 2013, p. 93).

This defeatist remark points to the recent unsettledness of trained radiologists. At stake are their very own professional expertise and the discipline's relation towards patient-centered health care. Since imaging techniques such as computed tomography (CT) or magnetic resonance imaging (MRI) are no longer exclusively applied and interpreted by radiologists and more and more therapeutic interventions are being guided by radiological imaging, radiologists are scared to lose their disciplinary expertise.[1] On a technological level, the future of radiology may seem bright but the expertise of radiologists is being questioned. In particular, professional associations of radiology, such as the American College of Radiology or the German Roentgen Ray Society, are emphatically planning political action that should act back on local routines. Campaigns such as "Imaging 3.0" are indicative of the search for an innovative self-conception of a clinical discipline in conversation with other medical specialties or experts from fields outside clinical

---

[1]The term "radiology" is used to describe a medical discipline that forms a "thought collective" (Fleck 1979 [1935], p. 39) sharing certain knowledge, socio-practical conventions as well as discourses. Radiology as a discipline is acted out according to local and epistemic contexts but there is a constant "communication of thoughts" (Fleck 1986 [1936], p. 103) between local instances and professional associations or other medical disciplines. Radiology in itself is a quite diverse field with multiple sub-specializations and paradigmatic orientations. Great epistemic divides are at stake even between diagnostic radiology and interventional radiology. Hence, empirical observations, specialist literature and discourses within the discipline need to be taken into account to detail and specify analytical questions and issues as well as contextually reassess the developments.

contexts. But what may seem to be a profound and proactive attempt to open up the discipline's clinical focus is another epistemic crisis of radiology induced by technological innovations.

Previously, another major technological and media-based turnover initiated the quest for a new radiological self-esteem about 15 years ago. At the time, the transformation of analogue visualization techniques and diagnostic workflows into all-digital environments challenged the radiological modes of processing and reading medical images. This so-called analogue-digital migration transformed the diagnostic process into a fully digital one. Print outs of radiological images or paper-based patient records were slowly but steadily integrated in digital software systems that process and visualize data from different imaging modalities such as CT, MRI and conventional X-ray (CR). The workflow of radiologists and socio-cultural interactions were both "relocated" in an "all-digital department" (Thrall 2005, p. 382).

These technological developments are inseparably connected with multifaceted institutional and socio-political processes on local, national and international levels. Interestingly, professional associations such as the American College of Radiology (ACR) try to frame the disciplinary challenges that are posed by technological innovations as a straightforward "evolutionary process". The ACR discursively disjoins different stages of technological innovation to launch new campaigns such as Imaging 3.0 and to claim constitutive turning points in radio-logical expertise and practice.[2] In particular, the use of established and novel imaging techniques and visualization technologies in other medical disciplines like radiation oncology or neurosurgery causes strategic and political campaigns to claim radiology's very own professional position in changing clinical environ-ments and in health care policy decisions. Radiology tries to demand back its leading role as a clinical discipline with long lasting traditions of visual expertise by simultaneously opening to new political demands such as patient-centered

---

[2]Radiologists who are involved in the recent debates about Imaging 3.0 use the "versions" of imaging as blueprints for different and future collective mind-sets in the course of technological developments (Ellenbogen 2013, p. 229). Besides, Imaging 2.0 is a trademark coined by the international company Philips Healthcare in the style of Web 2.0 to market their digital communication systems for radiological departments. As an economically driven campaign for the annual congress of the Radiological Society of North America in 2013, Philips' Imaging 2.0 is designed for "transforming care, together. Together, we are changing the expectations of what imaging is, how it should work, and what it can do today and tomorrow" (http://www.healthcare.philips.com/us_en/about/events/rsna/index.html, accessed 12 Sept. 2013). In this respect, it foreshadows the most recent campaign Imaging 3.0, which is driven in contrast by radiological associations.

health care (Goldsmith 2011). As technological possibilities become increasingly accessible and therapeutically valuable for other medical disciplines, radiology's very own expertise is arguably at stake. How are technological innovations mutually challenging disciplinary paradigms and specialties as well as the (self-) conception of possessing a certain expertise?

I will explore the professional "crises" and proclaimed culture shifts of radiology in the course of the analogue-digital migration as well as the Imaging 3.0 campaign by focusing on the relations between conceptual developments on a more macroscopic level and their impact on everyday practices within radiological settings.[3]

Concerning the context of the analogue-digital migration in the late 1990s, technological innovations such as digital image processing led to interdisciplinary negotiations between radiology and, in particular, software design on how to conceptualize a discipline and its novel digitally-based work practices. The design process of, for example, the international software vendor Agfa HealthCare was grounded on persona-based modeling, which is a tool to technologically codify the work practices of a certain user group. So-called personae are conceptualized as archetypical users based on empirical studies in the field to represent radiological workflows and expertise throughout the software design process. This raises questions of how a certain disciplinary knowledge and practice, such as radiology's, is anticipated by disciplines such as software design and further encoded and negotiated within software systems. The analysis of such modeling processes and of its implementation in everyday work routines will demonstrate how certain habits and social conventions cannot simply be integrated into automated modes of diagnosing.

The most recent technological innovations of image-guided interventions and their "disciplinary threat" to radiology will be discussed according to the nation-wide Imaging 3.0 campaign as well as to the locally acted out practices of tumor boards. The analysis of these gatherings allows me to illustrate the multi-disciplinary and patient-centered contexts of today's cancer treatment regimes and thereby points to the micropolitics of clinical medicine and the contested status of radiology as a discipline with a very specific visual expertise.

---

[3]Both foci are equally based on discourse analyses of subject-specific literature such as journal articles, news texts and handbooks, on observations in radiological and oncological departments at German university hospitals as well as on interviews with software experts from international health care companies. The samples and interviews draw on the media technologies under investigation, such as data processing or different visualization techniques.

2 "Imaging 2.0": The Analogue-Digital Migration and the Quest for Disciplinary Integration

Since around the year 2000, a fundamental technological and media-based change has been taking place in Western clinical radiology—the so-called analogue-digital migration, which is sometimes referred to as the analogue-digital conversion. The analogue workflows and modes of diagnosing, for example, reading printed scans on a light box or managing the clinical processes on the basis of paper-based patient records, was successively transferred to digital environments such as the picture archiving and communication system (PACS) (Dreyer et al. 2006). With the introduction of ever more technologically advanced imaging modalities such as innovative CT or MRI scanners, radiological expertise is being confronted with "data avalanches", ramified network architectures and advanced modes of visualization (Reiner et al. 2003). Regula Burri, in her ethnographic study of the clinical use of MRI, points out that there were "continuous efforts to accumulate symbolic capital" (Burri 2008, p. 41) within diagnostic radiology after the introduction of, in particular, magnetic resonance imaging in the 1980s. She identifies three contexts where these efforts became obvious, namely "the installation of the machines; the production and interpretation of the images; publication strategies and conference performances for the scientific community" (ibid). While the technological development of imaging could be read as an almost straightforward success story, discourse in the field of radiology rarely reflects social, cultural and epistemic factors. The boundaries of expertise are being destabilized not only in the clinical field. The disciplinary boundaries between physicians and information scientists are also becoming blurred. Burri concludes that the

> [...] diffusion and use of medical imaging technologies [...] occurred in the context of different epistemic, social and material reconfigurations. The clinical application of the new technologies induced a "crisis" in the "doxa" of radiologists, thus challenging their professional expertise and identity (Burri 2008, p. 55).

Therefore, the "crisis" of radiological expertise grows in size and impact by challenging developments and expertise outside the clinical context.

Before the migration from analogue to digital conventions of diagnostic practice, several radiological imaging modalities had of course already become digital, for example computed tomography (CT) or magnetic resonance imaging (MRI), but the circulation of data and images now became exclusively digital. The relocation of once material things and processes challenged the design of PACS software applications that enable radiologists to diagnose soft copies, i.e., data sets of medical visualizations, within a graphical user interface (GUI) with the help of various digital tools such as rulers or angle meters. Therefore, the analogue-digital migration causes far-reaching changes in the workflow and diagnostic practice of radiologists (Larsson et al. 2007; Tellioğlu and Wagner 2001). Perceived within the discipline as a "digital revolution we are now living" (Bryan 2003, p. 299), it overturns well-established conventions of thought and sight by transferring workflows into software systems.

2.1 "Prototyping Radiologists": How Programmers Anticipate a Profession

With the migration from analogue to digital workflows, software became a significantly powerful epistemic and operational tool (cf. Berry 2011, p. 3). Accordingly, questions such as how socio-technological connections and differences are constituted by the design and application of software arise.

The radiologists' light box with printouts of images on radiographic film still exists in some radiological departments but it is no longer used for diagnostic purposes. The new epistemic and aesthetic "arena" of visual expertise is a picture archiving and communication system (PACS), which is available from several vendors. To explore the challenges and the possibilities of the analogue-digital migration for radiology, it is important to trace the role of the software vendors who played a leading role in specifying the needs of the discipline by encoding and representing them in software solutions.[4]

Programming a software system, in particular its graphical user interface, is a complex endeavor as such but designing it according to already existing modes of thinking and perceiving without causing too much irritation is even more difficult.[5] Software systems do not only couple the user and computer, but also define "the

---

[4]The following paragraphs on persona-based modeling and the Agfa PAC System are based on interviews with three Agfa software designers, the analysis of design handbooks named by them, and the analysis of marketing material for the PACS published by Agfa HealthCare.

[5]I will use the notion of designer here to refer to human-computer interaction design that covers aspects of both programming and visual design. The division of labor between both specialties ranges according to the size and structure of the software company or complexity of the software system (cf., e.g., Coopmans 2011). In international health care companies such as Agfa or Siemens, the organizational unit of innovative responsibilities is a team. Focusing on one very specific aspect of a software creation process, the relations between different teams or silos sometimes might be quite loose as is the relation between programmers and designers.

interaction between the designer and the user *through* the system" (Dourish 2001, p. 56, italics in original). Designers determine formal processes of algorithms, data streams and computation in general, yet more strikingly they translate thought processes "into a mechanical process of action" (Trogemann 2010, p. 43). These thought processes are abstracted to machine code, but they derive from and, at another level, retroact on a material, cultural and social world that is evidently affected (Trogemann 2010, p. 44).

Several international health care companies like Agfa, General Electrics and Siemens offer software products to integrate, process, visualize, and archive image data from different imaging modalities. To simultaneously create and mirror the changing radiological workflows that are no longer based on printouts of images but on interactive visualizations of the human body, health care companies need to get an idea of a prototypical radiologist, his or her workplace and workflow, and individual as well as collective thought processes (Mayhew 1999, pp. 67–68). How do programmers prototype radiological expertise and its application by designing software systems such as a picture archiving and communication system? What is the conceptual anticipation of a radiological diagnostic workflow in the light of a media-based transformation that mutually causes epistemic as well as aesthetic innovations and—at the same time—constraints?

2.2 On Radiological Personas

To get an idea of a prototypical radiologist, Agfa HealthCare, for example, employs persona-based modeling, which is conceived as an "effective design and communication tool" (Agfa HealthCare 2007, p. 1). Initially, Agfa HealthCare commissioned the interaction designer Alan Cooper and his team to conduct empirical studies in radiological clinics to create specific personae.[6] Personae are not supposed to reflect "real people, but they represent them throughout the design

---

[6]"The personas used in IMPAX Agility have been drafted for Agfa HealthCare by the team of Alan Cooper. He created the Goal-Directed Design methodology and pioneered the use of personas as practical interaction design tools to create high-tech products. His team interviewed stakeholders at Agfa in Waterloo, Ontario, and product managers in Milwaukee, Wisconsin. Then they visited hospitals across North America and spoke with radiologists, PACS administrators, radiology IT staff, and Agfa services representatives. They came to Europe and conducted interviews in Italian, Belgian, Dutch and German hospitals. Most of the interviews took place in radiology reading rooms, making it easier to discuss common communication and information patterns, and identifying key behaviours in the reading and interpretation of radiology images" (Agfa HealthCare 2013, p. 2).

process. They are hypothetical archetypes of actual users. Although they are imaginary, they are defined with significant rigor and precision" (Cooper 1999, p. 124).

Since many software solutions in medicine are not developed in equal partic-ipation or work-sharing between health care companies and clinical specialists, software programmers need to get a valuable base for their "rigorous and precise" definitions of personae, which are, however, continuously adapted according to a specific clinic and its workflows, i.e., to the needs of the customer. Consequently, the use of software is conceptualized as "embedded into a set of practices" (Dourish 2001, p. 166) and needs to be approached by software specialists from outside the radiological realm. To define a rigorous and precise persona model and before the software itself couples and abstracts the interaction between the designer and the user, both have to meet in the context of application to exchange ideas about how software processes could be productively integrated. Therefore, pro-grammers or solution lifecycle managers conduct empirical studies to get a better understanding of habits and uses as well as of the workflows of the users. Using a contextual task analysis, designers try to

> [...] obtain a user-centered model of work as it is currently performed. That is, you want to understand how users currently think about, talk about, and do their work in their actual work environment (Mayhew 1999, pp. 67–68).

The prospective user is viewed as a prototype of real-world experience and not anticipated from a programming lab that is far from the contexts of application.[7] By being the object of empirical research, radiologists as prospective software users are challenged to reflect about their—often implicit—knowing and doing. Software designers use questionnaires or interviews as an empirical methodology to grasp the elusive factors of radiological knowing, such as visual expertise or an embodied knowledge of the interactional handling of radiological visualizations (cf. Alač 2011).[8]

---

[7]On early user modeling techniques in medical informatics with its "narrow conceptual world" that was "excluding as unscientific the informal, local information that could help them to design systems better suited to real users in particular workplaces", see Forsythe (2001 [1992], p. 11).

[8]User profiles are created according to "psychological characteristics (e.g. attitude, motivation), [...] knowledge and experience (e.g. typing skill, task experience), [...] job and task characteristics (e.g. frequency of user task structure), [...] physical characteristics (e.g. colour blindness)" (Mayhew 1999, p. 36). On the methodological tensions between ethnography and theory in human-computer interaction research, see Dourish (2006) and Bannon (2000).

But all these efforts can only be appropriations of a certain thought collective and thought style, particularly since socio-cultural routines, such as informal gatherings and individual workflows, need to be encoded into patterns that describe a persona and, finally, can be abstracted into machine code. What then is Agfa's "hypothetical archetype" of a clinical radiologist?

After Cooper and his team conducted empirical research at the start of the usability engineering lifecycle for Agfa's PACS in order to get to know the work routines, personal circumstances and interactional abilities, they created patterns that characterize the role of a typical user.[9] Since many work environments are quite diverse and cover multiple roles and actors, the concept of a persona needs to adapt to this circumstance.[10] As illustrated in Fig. 1, Agfa HealthCare names eight persona models that are continuously updated to design and reengineer Agfa's picture archiving and communication system IMPAX Agility.

One of the foundational personae of IMPAX Agility is constructed to represent an archetypical radiologist named Peter. Peter is supposed to be a white, middle-aged, friendly male, whose core activities are interpreting images and consulting. To anticipate the main goals of radiologists during the design process, the description of the Peter persona refers to one of his personal goals as to "feel effective, productive and confident" (see Fig. 1). This feeling, which is anticipated by Agfa HealthCare to be a need of many radiologists, must be engendered and mirrored by the software. Therefore, the graphical interface of IMPAX Agility, for example, includes a visual hint to track how many exams the radiologist has reviewed per day and how many image sets still need to be reviewed and diagnosed. The workload indicated by stacks of paper-based patient records and film-strips at the light box, which created a feeling of excessive demands or productivity, is nowadays reduced to a quantitative fact.

---

[9]Mayhew (1999) explains the various steps and tasks of a (conceptual) usability engineering life cycle. From a practitioner's point of view, she remarks that "[a]s far as the users are concerned, the user interface *is* the product. Just about their entire experience with the product is their experience with its user interface" (Mayhew 1999, p. 1). According to usability, the main question is how "*easy to learn* the user interface is for novice and casual users. Another is how *easy to use* (efficient, flexible, powerful) the user interface is for frequent and proficient users after they have mastered the initial learning of the interface" (Mayhew 1999, p. 1, italics in original).

[10]The information to update the existing persona profiles is gathered at "field trial sites, trade fairs and user-testing" because the life-world and in particular the technological world of personas evolves: "Peter, the radiologist, now has a tablet computer and a smart phone, he posts tweets on Twitter and updates his Facebook account. All of these didn't exist back in 2004 when Cooper described the persona" (Agfa HealthCare 2013, p. 3).

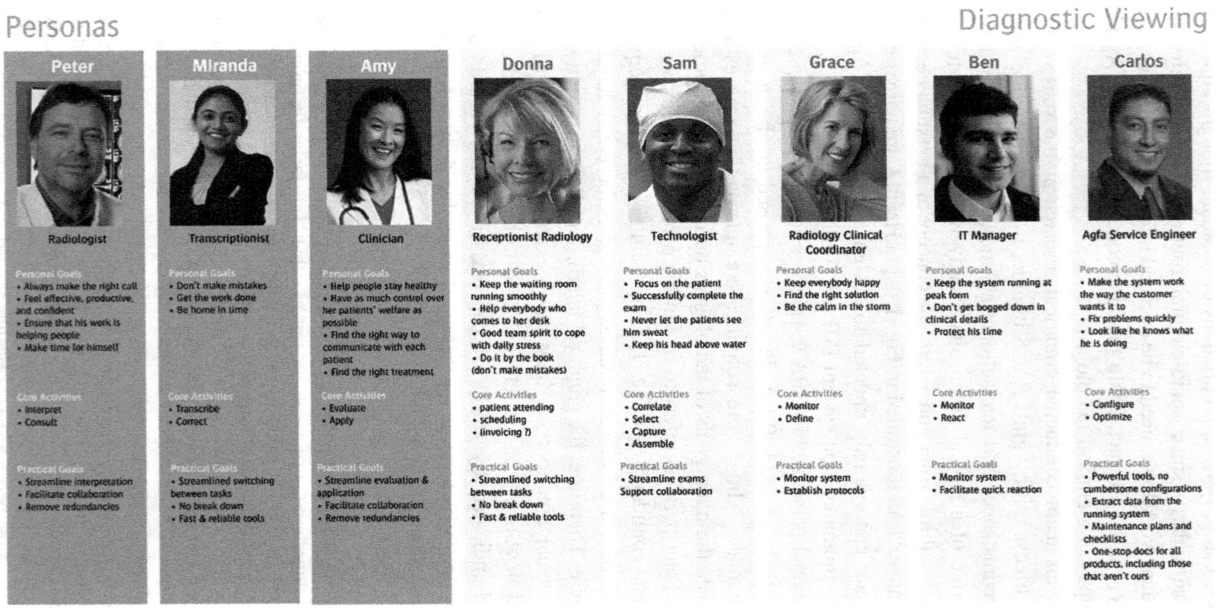

Personas

Diagnostic Viewing

Peter

Radiologist

Personal Goals
- Always make the right call
- Feel effective, productive, and confident
- Ensure that his work is helping people
- Make time for himself

Core Activities
- Interpret
- Consult

Practical Goals
- Streamline interpretation
- Facilitate collaboration
- Remove redundancies

Miranda

Transcriptionist

Personal Goals
- Don't make mistakes
- Get the work done
- Be home in time

Core Activities
- Transcribe
- Correct

Practical Goals
- Streamlined switching between tasks
- No break down
- Fast & reliable tools

Amy

Clinician

Personal Goals
- Help people stay healthy
- Have as much control over her patients' welfare as possible
- Find the right way to communicate with each patient
- Find the right treatment

Core Activities
- Evaluate
- Apply

Practical Goals
- Streamline evaluation & application
- Facilitate collaboration
- Remove redundancies

Donna

Receptionist Radiology

Personal Goals
- Keep the waiting room running smoothly
- Help everybody who comes to her desk
- Good team spirit to cope with daily stress
- Do it by the book (don't make mistakes)

Core Activities
- patient attending
- scheduling ?
- (invoicing ?)

Practical Goals
- Streamlined switching between tasks
- No break down
- Fast & reliable tools

Sam

Technologist

Personal Goals
- Focus on the patient
- Successfully complete the exam
- Never let the patients see him sweat
- Keep his head above water

Core Activities
- Correlate
- Select
- Capture
- Assemble

Practical Goals
- Streamline exams
- Support collaboration

Grace

Radiology Clinical Coordinator

Personal Goals
- Keep everybody happy
- Find the right solution
- Be the calm in the storm

Core Activities
- Monitor
- Define

Practical Goals
- Monitor system
- Establish protocols

Ben

IT Manager

Personal Goals
- Keep the system running at peak love
- Don't get bogged down in clinical details
- Protect his time

Core Activities
- Monitor
- React

Practical Goals
- Monitor system
- Facilitate quick reaction

Carlos

Agfa Service Engineer

Personal Goals
- Make the system work the way the customer wants it to
- Fix problems quickly
- Look like he knows what he is doing

Core Activities
- Configure
- Optimize

Practical Goals
- Powerful tools, no cumbersome configurations
- Extract data from the running system
- Maintenance plans and checklists
- One-stop-place for all products, including those that aren't ours

Fig. 1 Poster with the personae of IMPAX Agility (Agfa HealthCare 2013, p. 3)

But the anticipation of a radiologist's desired feelings hints at another aspect of the radiologist's persona in comparison to the other personae associated with the IMPAX software. The persona model of a radiologist mirrors the personal goal to "ensure that his work is helping people". In contrast, the persona of a clinician named Amy included as a personal goal her desire to "have as much control over her patients' welfare as possible". The persona of Amy includes other socio- professional features that need to be integrated into the software application. Clinicians are supposed to need radiological information about their patients to provide therapy, but not to interpret image data themselves. But a radiologist, conceptualized as the Peter persona, primarily communicates with other clinicians and physicians who need information about their patients. Hence, a software designer needs to integrate features which allow radiologists to focus on image interpretation inasmuch as the application needs to fulfil the needs of clinicians who just want quick and comprehensive information about the health status of their patients.

Even if the software system Agfa IMPAX Agility connects all the clinical subspecialties with one another, it fosters epistemic boundaries within clinical medicine as the roles of subspecialties are reinforced. Radiologists are conceptualized as physicians mainly "invisible" to patients and radiology as such rather as a commodity than a specialty within clinical contexts (Glazer and Ruiz-Wibbelsmann 2011, p. 19; cf. Star and Strauss 1999, pp. 23ff.). The abstraction from empirical data during the software design process, on the one hand, restabilizes the practical boundaries between radiological doing and "designerly ways of knowing" (Cross 2001) while on the other hand it embraces the "politics of design" (Bannon 1995) to regenerate disciplinary boundaries by technological means.

2.3 Digital Data Overload: Collective Negotiations of a "New Sight"

The efforts to constitute new disciplinary and technological standards for radiological workflows not only circumscribed the negotiations with commercial software designers but also took place in medical contexts, such as in professional associations. Radiological associations in North America and Germany identified the information and data overload generated by digital technologies and communication systems as one of the most pressing, but equally beneficial issues in the course of the analogue-digital migration (Andriole et al. 2004).

The benefits of PAC systems are clear. Within seconds after an image is acquired, it can be viewed by the radiologist and any number of referring and treating physicians simultaneously. There is no film to be lost or stolen. CT exams with a thousand images are becoming common and simply cannot be managed effectively on film (Hirschorn 2006, p. 3)

In everyday radiological practice, these promises of an overall digitalization such as operational and material flexibility, time efficiency and expansion of visualization modes were eroded by "data avalanches".[11] The enthusiasm about the substantial changes cannot hide the fact that not only technological and diagnostic standards needed to be reconsidered, often in exchange with software experts and health care companies, but also social and communicative structures within radiological departments were transformed. The adjustment of radiologists to new modes of diagnosing covered a lot more, and often intangible factors than the increase in speed and information.

One example is the tremendous change in workflows in all-digital radiology departments. Besides the technological and legal requirements, such as data exchange standards, the integration of the Agfa IMPAX software into the everyday work routines of radiological departments challenged, amongst other things, the conventions of diagnostic practices. The increased possibilities resulting from the new technology paradoxically limited the application of knowledge since many radiologists needed to become aquatinted with the software that contested their individually and collectively established diagnostic conventions. These conventions were now confronted with the "realms of possibility" that the software offered (Dodge and Kitchin 2011, p. ix; Manovich 2008). The examination of filmstrips with tools like spacers or angle meters, a very haptic procedure, was delegated to several clicks within the graphical user interface (GUI) of a PACS (Krupinski and Kallergi 2007, p. 667). The next-to-last release of the Agfa HealthCare PAC system included the possibility to create a personalized log-in profile for each radiologist in the department. Radiologists were encouraged to select among more than 119 menu items and 87 different tools to define their own profile. The software included this number of prefigured digital tools and configurations, and all the user had to do was to enable the functions so that they were displayed automatically during every log-in. Very few radiologists used the personal configurations. They could not relate to all these possibilities because many of them use 10–15 software tools routinely and, paradoxically, the various options offered by the software limited the interaction with the software. The possibilities to configure a

---

[11]Andriole et al. refer to an informal study of the Department of Radiology at the Mayo Clinic Jacksonville, Florida, which "determined that roughly 1500 images were generated and stored in 1994. In that same practice in 2002, an average of 16,000 images were acquired each day" (Andriole et al. 2004, p. 237).

personal profile have been removed from the newer version of the software, once again limiting the options for personalization.[12]

In addition, the sequence in which the images are diagnosed needs to be adapted to the new technology. An experienced radiologist reported that he needed to change his working procedures in accordance with the software's display modalities. The system first of all supplies and displays an overview of the body region that was scanned (topogram), followed by scans in greater detail and more slices because it takes a while for the image reconstructions to be computed. Filmstrips, once they were hung up, could be viewed synchronically, while the radiologist now adapts to the software conventions.[13]

These conventions also "tie" the radiologist to a computer screen. Rarely more than one person is sitting in front of a digital workstation (Krupinski and Kallergi 2007). But even though a series of images can be viewed simultaneously on several workstations, radiologists must occasionally meet physically to discuss their findings. Even if the software designers tried to match the embodied analogue diagnosing conventions of the radiologist, they could not add and digitize the "surplus" feature of face-to-face interaction and physical social meetings.

The increase of digital data gathered by more powerful imaging modalities and by functional imaging, the time pressure resulting from digital communication tools and innovative aesthetic features offered by software solutions were, however, causing deep worries within the discipline. In particular, the contrast between the number of diagnostic images and the available time evoked the concern that more medical errors could result. Besides, economic pressure and the shortage of well-trained radiologists tightened the "information overload" per radiologist (Bryan 2003).

---

[12]This practice was mentioned by senior radiologists as well as by younger professionals during my observations and interviews in a radiological department of a German university hospital. Additionally, a software designer of Agfa HealthCare explained in an interview, that the personal log-in feature was not integrated into newer releases of the software since many users would not use it.

[13]Radiological images, such as computed tomography scans, were printed out as matrixes of a certain number of scans on one film slide. Each transversal scan of a body region was printed out one after another so that the radiologist could see the first and the last "slice" of, e.g., the thorax on one filmstrip (depending on the total number of slices gathered and reconstructed during one scan). As the author observed in clinical settings, the visualization of image data by software applications is mainly based on interactive operations such as scrolling to browse through a "stack" of tomographic slices one after another. This has other diagnostic and aesthetic advantages but initially challenged the radiologists' modes of image interpretation.

The subsequent international initiatives of the radiological community like Integrating the Healthcare Enterprise (IHE) and Transforming the Radiological Interpretation Process (TRIP) have stemmed from and fed back into local clinical contexts and individual workflows. Whereas IHE aimed at forming a technological basis for interoperable image data transfer among different imaging and network modalities, TRIP pointed directly to the changing modes of image perception due to digital softcopy reading (Fridell et al. 2007). The discussion about the profession's self-conception was outlined by a paper from the TRIP committee in 2004 to "address the problem of information and image data overload" and to "foster interdisciplinary research on technological, environmental and human factors to better manage and exploit the massive amounts of data" (Andriole et al. 2004, p. 235).[14] Aimed at a "paradigm shift in the radiological interpretation process" (ibid.), a group of radiologists named the challenges to their professional work that were caused by the technological changes of the analogue-digital migration. The paper's title, "Addressing the Coming Radiology Crisis", evokes a certain sense of self-critique on the one hand but on the other hand signals the discipline's own capacities to deal with a crisis by discussing it early and having

> [...] a wonderful opportunity to change the radiological interpretation process, improving both the quality of patient care and the efficiency for future radiologists and their electronic practice (Andriole et al. 2004, p. 236).

The TRIP initiative named both technology and human perception as objectives of reconsideration, as did other subject-specific literature (cf. Siegel et al. 2006). But whereas suggestions outlined how technological applications and hardware could be better standardized, it is still not quite obvious how the problem of limited human image perception should be dealt with. Some of the main issues in radiological image interpretation are the problem of sufficient training, the articulation of implicit knowledge, and guidelines for image inspection. Therefore, further collaboration with software designers and programmers as well as with "creative thinkers from a number of fields" (Andriole et al. 2004, p. 238) was anticipated by major professional societies in order to tackle the problem of data and information overload. Whereas this is a problem of diagnostic productivity and volume, most recent debates within radiological associations have focused intensely on radiology's role in patient-centered health care. The campaign Imaging 3.0 heralds a new disciplinary crisis that is induced by the further technological innovations but also by debates within clinical medicine. In particular, radiology's clinical status as a discipline of diagnostic visual expertise is being contested by a trend to employ

---

[14]The TRIP committee is part of the Society for Computer Applications in Radiology.

imaging techniques and images for therapeutic purposes in various medical disciplines.

3 "Imaging 3.0": The Rise of Image-Guided Therapies and the Recent Crisis of Radiology

One of the most recent topics concerning the disciplinary self-conception and specialty of radiology is the application of radiological imaging techniques by other medical disciplines and not only for diagnostic but also for therapeutic and interventional purposes. Image data gathered by technologies such as CT and MRI are visualized and reconfigured to diagnose, plan and perform medical treatment and surgery. For about five years, the disciplinary boundaries of radiology are again being contested by the increasing clinical application of image-guided interventions and their technological innovations.

These image-guided interventions are not only changing the way medical images are perceived and used but also destabilize radiology's fields of competency within clinical contexts. Radiologists are no longer the predominate experts in producing, reading and handling radiological images. Surgeons, radiation oncologists and internists are becoming more and more accomplished in diagnostically viewing pre-operative CT or MRI images as well as using software to manipulate them in the course of treatment planning. The therapeutic use of radiation and other image-guided techniques to treat illnesses such as cancer is questioning the exclusive status of radiologists as the experts for image- and radiation-based procedures in clinical contexts. Besides, the shift from diagnostic to "therapeutic" imaging triggers the self-conception of diagnostic radiology. Even if radiologists can specialize in the course of their clinical training in interventional radiology, a practice that includes diagnosing and minimally invasively treating diseases on the basis of radiological imaging techniques, this does not necessarily include the responsibility for a full-circle patient treatment based on imaging modalities.[15]

Hence, other clinical experts are appropriating the technologies, visualization tools and iconic knowledge of radiology according to their specialty and therapeutic

---

[15]The regulation for further education released by, for example, the Berlin Medical Association includes in the section on specialization in radiology only minor hints regarding qualification in direct patient supervision. In contrast, the section on specialization in internal medicine stresses the necessity of social and communicative competence (Ärtztekammer Berlin 2012, pp 50ff.).

field. Of course, radiologists are not excluded from these developments but radiology as a clinical profession is challenged to adapt to "shifting cultures" (Muroff 2013) within medicine.

I will explore these very recent socio-cultural and epistemic changes by looking at two instances. The above-mentioned nationwide campaign Imaging 3.0 draws attention to the more macroscopic reconsiderations and active restructuring of a clinical discipline in relation to others. The second focus will be on a microscopic level, describing the example of so-called tumor board reviews. These weekly conferences involve several clinical experts, among them surgical and radiation oncologists, pathologists and radiologists, to discuss and plan the treatment options of a cancer patient. While tumor boards are a legal necessity, their concrete "scenario" and interdisciplinary infighting can only be observed in clinical microcontexts.

3.1 "Radiology Is Getting an Upgrade": The Imaging 3.0 Campaign

Very recently, the American College of Radiology (ACR) launched an image campaign—almost to be taken literally—called Imaging 3.0 (Ellenbogen 2013). After mastering the demands of the disciplinary and epistemic changes induced by the analogue-digital migration, radiology now faces new technological and structural changes. The discipline's specialty and exclusive status has become even more contested by the expansion of image-guided interventions and therapies.

Image-guided therapies are highly complex forms of treatment based on and guided by various imaging techniques. The most common uses are image-guided radiation therapy in cancer treatment or minimally invasive image-guided thoracic and neurosurgery. Therefore, imaging technologies are used to diagnose and pre-plan the interventions as well as to intra-operatively track and monitor the treatment (Helmberger et al. 2013, p. 2).

Imaging technologies are obviously being used to carry out patient-centered treatments, such as cancer radiation therapy (Jaffray 2012), and no longer "just" to diagnose a certain illness. Intrinsically related to this development is the entanglement of different clinical specialties that are producing, interpreting and employing radiological images according to their specific therapeutic interests. The radiologists' differential diagnosis is negotiated within an interdisciplinary medical team, such as oncologists and surgeons, and according to the treatment undertaken for a patient. Radiologists are supposed to leave their dark reading rooms of interpretation and approach patients and referring physicians with their very own visual expertise and a specific treatment proposal ready to hand. "However, the impact of radiology on the regulatory medico-legal, technical and radioprotection issues in this field have [sic] not yet been defined" (Helmberger et al. 2013, p. 3). This very recent development caused unsettledness within the discipline regarding everyday radiological practices and within professional institutions.

The ACR and some of its members act as precursors to initiate a disciplinary rethinking and to secure the authority of radiological expertise. For example, radiology professor Lawrence Muroff insinuates that his colleagues have a certain apathy, which

> [...] has caused radiologists to take a backseat to the many other interests that vie for their limited available time. This [...] makes it difficult to get radiologists to understand that current socioeconomic and competitive forces mandate proactive thinking and action if radiologists are to enjoy the quality of life and financial benefits to which they have become accustomed (Muroff 2013, p. 93).

The ACR hence encourages "shifting cultures" of contemporary radiological practice by launching the initiative Imaging 3.0. Paul Ellenbogen, Chair of the ACR Board of Chancellors, acts as a spokesperson:

> [...] Imaging 3.0 is a call to action for radiologists, policymakers, payers, referring physicians, and patients to provide optimal imaging care from the moment a clinician considers ordering an imaging study or treatment until that referring physician receives and understands an actionable report with evidence-based recommendations. The goal is easy to put into words: To deliver all the imaging care that is beneficial and necessary and none that is not (Ellenbogen 2013, p. 229).

Radiologists should get more involved in patient care and treatment instead of providing services for other clinical disciplines (Levin et al. 2013, p. 647). The slogan is to "maximize radiologists' value":

> We need to seek greater collaboration with other physicians, we need to empower patients, and, not unimportantly, we need to change the discussion in Washington (Ellenbogen 2013, p. 229).

The profession seems to be in deep concern about its very own disciplinary status and clinical impact since its authority to produce, read and interpret radiological images is being questioned by the innovation and circulation of imaging technologies.[16]

---

[16]In Germany, similar campaigns and developments can be observed. For example, Prof. Michael Forsting, the former president of the German Roentgen Ray Society (Deutsche Röntgengesellschaft e.V.), entitled his keynote address at the latest annual congress

Notably, the changing clinical and economic regimes mean that a radiologist's very own visual expertise does not seem sufficient anymore. Questions of the appropriateness of radiological exams ordered by referring physicians or the communication of results to patients come into play. Both aspects refer to the assumption that clinical knowledge and the integration of radiologists might not be "valuable" enough while their professional efficiency is at a high volume (Levin et al. 2013, p. 650).

Hence, radiologists are called to action to adapt their disciplinary, epistemic and even personal boundaries to provide patient-care instead of solely being responsible for imaging services with profound expertise. The discipline's self-conception is taken "beyond imaging" and therefore is also challenging new modes of socio-cultural interaction. "Radiologists must integrate themselves into the medical, social, and political fabrics of their hospitals and their communities [...]" (Howell 2013). In so doing, the Imaging 3.0 initiative promises that "radiology is getting an upgrade" (Howell 2013) which will shift the socio-epistemic culture and the certain expertise of radiology from "volume-based service to concentrating on value-based practice" (Howell 2013). One of the pragmatic settings where radiologists need to integrate and develop new forms of interaction as well as to make epistemic and economic claims is the weekly meeting of tumor boards.[17]

(Footnote 16 continued)
"Radiology is the Future—Radiology of the Future" (Radiologie ist Zukunft—Radiologie der Zukunft). Additionally, a poster campaign under the heading "Medicine with Perspective" (Medizin mit Durchblick) was launched. Radiological visualizations on the posters are accompanied by questions addressing the visual knowledge of the spectator in contrast to radiological expertise. One of the questions that goes along with a greyscale image of a part of the spine bone is: "You see a bread for toasting? We see a slipped disc.—Discover the fascinating world of radiology and nuclear medicine".

[17]According to the role of associations of radiology some positions in the discourse strongly recommend that it "is important for radiology organizations to draft position statements when turf battles arise that affect their members. These statements should represent the collective viewpoint and ideas of organization and, once adopted, be distributed to the membership. That way, if and when confrontations occurred at members' hospitals, they will be armed with some ammunition for the debate" (Levin et al. 1999, p. 304). This top-down model of forming a disciplinary consensus that can be "acted out" in local contexts somehow mirrors the clinical hierarchies that are relying on rigid reputational and professional statuses and communication habits.

3.2 Tumor Boards: Multidisciplinary (Visual) Expertise

The complexity of cancer and its variations require multiple experts from different medical subdisciplines. As a legal requirement, multidisciplinary tumor board reviews are an essential feature for planning a patient's treatment. Different tumor boards for specific clinical indications are installed in clinical routines, for example a thorax tumor board particularly for patients suffering from lung cancer (Gould 2011). Physicians with a background in clinical oncology, internal medicine, radiation oncology, surgery, radiology, and pathology meet in person or via videoconferences, depending on the size and internal structures of a clinic. The physician who is responsible for the care of the patient at his or her ward describes the personal circumstances, symptoms, clinical indication, and previous treatments and, in some cases, proposed further treatments to his or her colleagues.[18]

The patient is "visible" through his electronic patient record and most prominently in radiological images. Prior images of the patient gathered by, for example, CT or MRI measure and visualize the inner characteristics of the human body, such as anatomical structures and biochemical processes. The imaging techniques and machines are not necessarily installed in the radiological department because radiation oncology clinics might also own a computed tomography scanner. But since the images are acquired and processed digitally via network systems they can be viewed throughout the different departments. In tumor boards, radiologists report the results of their examinations while scrolling through image stacks to give everyone a "picture" of the patient's current status. Interestingly, in most cases, the radiologist is still the expert for reading and interpreting the visualizations but the other clinical experts will most likely discuss further treatment options and at the end decide what to do. Radiological expertise in this process is not integrated into the decision-making process for patient-centered therapies but rather conceived as a specialty that reveals the visually encoded features of an illness on the basis of extensive training and experience. Other physicians do not suppose this expertise to rely on experience and knowledge of concrete clinical symptoms and treatment options since radiology is supposed to be a discipline that is mostly "invisible" to patients and hence has no extensive experience in treating them directly (Glazer and Ruiz-Wibbelsmann 2011, p. 20). Additionally, other clinical disciplines such as radiation oncologists are gaining more and more visual expertise since they themselves possess imaging technologies and need to consult images as a basis for making diagnoses and conducting treatment. Furthermore, in negotiation with

---

[18]The following section is based on observations in a German university clinic during different tumor boards.

other medical disciplines about who will treat the patient by which methods, diagnostic radiology often has not much to offer because patient care has until now not been within its epistemic and practical boundaries (Levin et al. 2013, p. 648).[19]

In the routine practice of radiology in a clinical setting, the propagated paradigm shift of radiology from a discipline relying mainly on diagnostic interpretation towards one providing patient-centered health care has hardly been conceived. In particular, radiologists themselves see their role more as specialists for radiation protection or as trainers for other physicians conducting image-guided interventions instead of being close to patients (Helmberger et al. 2013, p. 6; Levin et al. 1999, p. 302).

4 Conclusion

Technological innovations in the clinical context have caused unsettledness in the discipline of radiology and concerning its socio-epistemic status. In everyday clinical work routines, these developments challenge the self-conception of radiologists as visual experts and their relationship towards patient care. On another level, professional associations of radiology launch public and political campaigns of disciplinary "reinvention" to testify both to radiology's changing visual expertise and to its new role in patient care.

The analogue-digital migration posed a challenge to the visual expertise of radiologists and to work routines in radiology not only because of the profound technological shift that the digitalization of image processing initiated but also because the migration from analogue to digital visualization modes was accompanied by an inevitable collaboration with software experts. In particular, innovative technology such as picture archiving and communication systems could not be designed and installed without "prototyping" the prospective users by persona-based modeling. This means of designing human-computer interaction abstracts pragmatic scenarios into different levels of machine code. Reassessing these abstractions in everyday applications makes it obvious how socio-cultural routines cannot be covered by software applications. The more operational and aesthetic options that software and its interfaces offer, the more the latter tend to

---

[19]Helmberger et al. presume that having radiologists perform image-guided interventions would subserve health care and economic benefits since radiologists do not tend to order tests or treatments on a patient that they themselves would perform or for which they would receive financial incentives: "IGT performed by radiologists is almost always free from problems related to self-referral. From a medical point of view, this independent position should be recognised and brought to the attention of healthcare providers" (Helmberger 2013, p. 5).

ignore the diagnostic routines and implicit knowledge of radiologists. Software designers standardized and quantified once material workflows within a software application on the assumption of output-oriented radiologists and according to legal restrictions. These standardizations were negotiated in clinical practice by social routines and at the same time constituted a certain "culture of discomfort". Hence, more campaigns by professional radiological associations were launched to proactively structure the dialogues between radiologists and programmers.

The most recent Imaging 3.0 campaign identifies and addresses the lines of demarcation between disciplines that run through clinical specialties. Radiologists and their professional expertise are evidently being challenged by other clinical disciplines such as radiation oncology and surgery. A practical and at the same time epistemological shift is taking place which questions both the "reign" of radiology's visual expertise and the status of radiological imaging as being either diagnostic or therapeutic. In particular, the growing installation and use of image-guided interventions raises the awareness of radiologists and their professional institutions that they need to get involved in epistemic and professional discussions about patient-centered health care.

> A culture shift for radiologists is not a luxury; it is an imperative for professional survival. Are you and your colleagues prepared to do what is necessary to survive and thrive in the turbulent times ahead? (Muroff 2013, p. 98).

The "culture shift" propagated by Muroff points towards a turn from diagnostic output towards therapeutic care within radiology. It is too early to determine whether this turn will go further than integrating radiologists as experts for radiation protection or technological innovations into the future development of image-guided interventions (Matsumoto et al. 2013, p. 100). Whatever happens, it will further challenge the visual expertise and the self-conceptions of radiologists, both in clinical settings and on the level of professional associations.

Acknowledgments My research was funded by the German Federal Ministry of Education and Research (01UB0925A) as well as by the DFG Cluster of Excellence 1027 "Image Knowledge Gestaltung. An Interdisciplinary Laboratory".

References

Ärztekammer Berlin (2012) Weiterbildungsordnung der Ärztekammer Berlin, version effective 17 March 2012. http://www.aerztekammer-berlin.de/10arzt/15_Aerztliche_Weiterbildung/10_wbo/00_WbO_2004_inkl_1_bis_9_Nachtrag.pdf. Accessed 15 March 2014

Agfa HealthCare (2007) IMPAX Insights #1—Persona-based workflow design. Supporting the workflow needs of individual users. http://www.agfahealthcare.com/he/global/en/internet/main/products_services/product-info/white_papers/persona_based_workflow_design.jsp. Accessed 28 Jan 2015

Agfa HealthCare (2013) IMPAX agility, White Paper—Persona based approach. http://www.agfahealthcare.com/global/en/main/resources/white_papers/index.jsp. Accessed 15 Aug 2013

Alač M (2011) Handling digital brains. A laboratory study of multimodal semiotic interaction. MIT, Cambridge

Andriole KP, SCAR TRIP Subcommittee (2004) Addressing the coming radiology crisis. The Society for Computer Applications in Radiology Transforming the Radiological Interpretation Process (TRIP™) Initiative. J Digit Imaging 17(4):235–243

Bannon L (1995) The politics of design: representing work. Commun ACM 38:66–68

Bannon L (2000) Situating workplace studies within the human-computer interaction field. Cambridge University Press, Cambridge

Berry DM (2011) The philosophy of software: code and mediation in the digital age. Palgrave Macmillan, London

Bryan RN (2003) The digital rEvolution. The millennial change in medical imaging. Radiology 229(2):299–304

Burri RV (2008) Doing distinctions. Boundary work and symbolic capital in radiology. Soc Stud Sci 38(1):35–62

Cooper A (1999) The inmates are running the asylum. Why hi-tech products drive us crazy and how to restore the sanity. Sams Publishing, Indianapolis

Coopmans C (2011) "Face value": new medical imaging software in commercial view. Soc Stud Sci 41(2):155–176

Cross N (2001) Designerly ways of knowing. Design discipline versus design science. Des Issues 17(3):49–55

Dodge M, Kitchin R (2011) Code/space: software and everyday life. MIT Press, Cambridge

Dourish P (2001) Where the action is. The foundations of embodied interaction. MIT Press, Cambridge

Dourish P (2006) Implications for design. Proceedings of the ACM conference human factors in computing systems. IEEE Computer Society Press, Los Alamitos, CA, pp 541–550

Dreyer KJ, Hirschorn DS, Thrall JH, Mehta A (eds) (2006) PACS. A guide to the digital revolution, 2nd edn. Springer, New York

Ellenbogen PH (2013) Imaging 3.0. What is it? J Am Coll Radiol 10(4):229

Fleck L (1979 [1935]) Genesis and development of a scientific fact. University of Chicago Press, Chicago

Fleck L (1986 [1936]) The problem of epistemology. In: Cohen RS, Schnelle T (eds) Cognition and fact. Materials on Ludwik Fleck. Reidel, Dordrecht, pp 79–113

Forsythe DE (2001 [1992]) Blaming the user in medical informatics. The cultural nature of scientific practice. In: Hess DJ (ed) Studying those who study us. An anthropologist in the world of artificial intelligence. Stanford University Press, Stanford, pp 1–16

Fridell K, Edgren L, Lindsköld L, Aspelin P, Lundberg N (2007) The impact of PACS on radiologists work practice. J Digit Imaging 20(4):411–421

Glazer GM, Ruiz-Wibbelsmann JA (2011) The invisible radiologist. Radiology 258(1): 18–22

Goldsmith J (2011) The future of radiology in the new health care paradigm. The Moreton Lecture. J Am Coll Radiol 8(3):159–163

Gould MK (2011) Multidisciplinary approach to the evaluation of the lung cancer patient. In: Kernstine KH, Reckamp KL (eds) Lung cancer. A multidisciplinary approach to diagnosis and management. Demos Medical Publishing, New York, pp 7–42

Helmberger T, Martí-Bonmatí L, Pereira P, Gillams A, Martínez J, Lammer J, Malagari K, Gangi A, de Baere T, Adam JE, Rasch C, Budach V, Reekers JA (2013) Radiologists' leading position in image-guided therapy. Insights Imaging 4(1):1–7

Hirschorn DS (2006) Introduction. In: Dreyer KJ, Hirschorn DS, Thrall JH, Mehta A (eds) PACS. A guide to the digital revolution, 2nd edn. Springer, New York, pp 3–6

Howell WLJ (2013) Imaging 3.0: Radiology gets a facelift. http://www.diagnosticimaging. com/practice-management/imaging-30-radiology-gets-facelift. Accessed 20 Sept 2013

Jaffray DA (2012) Image-guided radiotherapy: from current concept to future perspectives. Nature Rev Clin Oncol 9(12):688–699

Krupinski EA, Kallergi M (2007) Choosing a radiology workstation: technical and clinical considerations. Radiology 242(3):671–682

Larsson W, Aspelin P, Bergquist M et al (2007) The effects of PACS on radiographer's work practice. Radiography 13:235–240

Levin DC, Rao VM, Berlin J (2013) Ensuring the future of radiology: how to respond to the threats. J Am Coll Radiol 10(9):647–651

Levin DC, Rao VM, Bree RL, Neiman HL (1999) Turf battles in radiology: how the radiology community can collectively respond to the challenge. Radiology 211(2):301–305

Manovich L (2008) Software takes command. http://lab.softwarestudies.com/2008/11/ softbook.html. Accessed 8 Nov 2011

Matsumoto AH, Adams MJ, Bello JA, Lozano K, Rosenthal SA, Swan TL (2013) Commentary on "culture shift"—radiologists and radiation oncologists adding value to the health care system. J Am Coll Radiol 10(2):99–100

Mayhew DJ (1999) The usability engineering lifecycle. A practitioner's handbook for user interface design. Morgan and Kaufmann Publishers, San Francisco

Muroff LR (2013) Culture shift: an imperative for future survival. J Am Coll Radiol 10 (2):93–98

Reiner BI, Siegel EL, Siddiqui K (2003) Evolution of the digital revolution: a radiologist perspective. J Digit Imaging 16(4):324–330

Siegel EL, Reiner BI, Knight N (2006) Reengineering workflow. The radiologist's perspective. In: Dreyer KJ, Hirschorn DS, Thrall JH, Mehta A (eds) PACS. A guide to the digital revolution, 2nd edn. Springer, New York, pp 97–123

Star SL, Strauss A (1999) Layers of silence, arenas of voice: the ecology of visible and invisible work. Comp Supp Coop Work 8(1):9–30

Tellioğlu H, Wagner I (2001) Work practices surrounding PACS. The politics of space in hospitals. Comp Supp Coop Work 10(2):163–188

Thrall JH (2005) Reinventing radiology in the digital age, part I. The all-digital department. Radiology 236(2):382–385

Trogemann G (2010) Code and machine. In: Gleiniger A, Vrachliotis G (eds) Code: between operation and narration. Birkhäuser, Basel, pp 41–54

The Use and Influence of Indicators in Decisions About Technological Innovation

Quantitative Results from Questionnaires in Portugal

Nuno Boavida

Abstract

This work focuses on the role that indicators play in decisions about technological innovation in three knowledge-intensive groups: Researchers in public institutions, leaders in research and development in industry, and policy-makers. It presents quantitative results obtained from three questionnaires sent to these groups in Portugal. The study suggests that indicators are instruments in decisions about technological innovation but that their influence depends on the social context and the type of decision. Results show that indicators were used in all the groups but were not significantly influential in decisions on technological innovation. Researchers were more influenced by indicators than by their social relations, revealing a balance between an instrumental use, a symbolic use and no use at all of indicators. Those in this group focused their decisions on the acquisition of products or technology, and identified its main influence as being in the sources and users of knowledge. The majority of the business and policy-makers revealed that indicators were mostly used in a symbolic way, and that they were more influenced by social relations than by indicators. Those in the business group focused their decisions on the development of products or technology, and declared that hierarchies and users exerted a stronger influence on their decisions. The policy-makers focused their decisions on the design of innovation policies, and they too were influenced more strongly by hierarchies and knowledge sources.

N. Boavida (✉)
Karlsruhe, Germany
e-mail: nuno.boavida@kit.edu

© Springer Fachmedien Wiesbaden GmbH 2016   59
B.-J. Krings et al. (eds.), *Scientific Knowledge and the Transgression of Boundaries,* Technikzukünfte, Wissenschaft und Gesellschaft / Futures of Technology, Science and Society, DOI 10.1007/978-3-658-14449-4_4

Keywords
Technology innovation · Indicators · Evidence · Decision-making ·
Policy-making · Technology assessment

1   Introduction

There has been an increase in the *production* of quantified measures in most societies in the last century.[1] The use of these measures can be observed in very different fields, such as in companies, government administration, military activities, and science (Porter 1995). There is a general consensus that this quantification produces indicators useful for humans to make coherent and enhanced decisions. Although such indicators are considered a useful instrument for guiding and improving decision-making, researchers have given them relatively little attention in understanding their relevance. In a policy context, a few studies revealed that indicators were ignored and their impact limited (see Porter 1995; Gudmundsson and Sørensen 2012; Freeman 1995; Power 1997; Grupp and Schubert 2010; Grupp and Mogee 2004). In the context of innovation, only a few studies have provided evidence of their concrete use and influence on decision-making (see Gudmundsson and Sørensen 2012; Fioramonti 2014; Boavida et al. 2013). This paper reveals that, although indicators are described as instruments for decision-making, their final influence in a knowledge-intensive context is significantly dependent on the social context and the type of decision.

The paper addresses in a quantitative way the question as to how members of knowledge-intensive groups involved in technological innovation make their decisions. It seeks to understand the possible discrepancy between the consensus about the usefulness of indicators and the reality of their use and influence in a knowledge context. This work focuses on knowledge-intensive groups because they are the most prone to use quantified measures in their work. In fact, members of knowledge-intensive groups can use these measures more than those in other groups because, in their work, they have facilitated access to knowledge, information, and indicators. The paper studies the behavior of knowledge-intensive groups that are engaged in decision-making related to technological innovation because these types of decisions are concrete, identifiable, knowledge-dependent,

---

[1]See, for example, the prolific data-gathering work about the world economy of Maddison (2001, 2003).

and relevant from a socio-economic perspective. Three groups were analyzed, namely: (1) a researchers group composed of public researchers, academics, and research and development (R&D) health personnel; (2) a business group composed of companies' research, development and innovation (R&D&I) personnel;[2] and (3) policy-makers associated with technological innovation. The research combined a study of the literature, analysis of official documents, three questionnaires, and interviews and was conducted in Portugal.

The rest of this paper is divided in four parts. Section 2 discusses the influence of indicators in the context of decisions regarding technological innovation. Section 3 describes the methodology used in the work, focusing in more detail on the construction of the samples. Section 4 presents quantitative results regarding whether indicators were used and the influence they had on decision-making about technology. It starts by distinguishing in a quantitative way the use of indicators from their real influence, analyzing afterwards the influence of the social context and the type of technology involved. The last section discusses the conclusions presented in the preceding sections, and elaborates on future lines of research for this topic.

2 Decision-Making and Indicators Regarding Technological Innovation

In this section, I introduce the framework employed to study the presence of indicators in the context of technological innovation. After identifying the main knowledge gaps in the literature, I argue that the study needs to distinguish the use of indicators from their real influence, detect contextual influences, and identify the types of technology decision involved.

2.1 Decision-Making in the Context of Innovation

An *innovation* is commonly defined as a new idea, device or method. The result of an innovation is normally a new technological product or service or the significant improvement of an existing product or process. According to Utterback (1974), innovation is distinct from an invention or technical prototype, and refers to

---

[2]In order to simplify the presentation of the data, the researchers group will be briefly denominated as researchers, and the group of business R&D&I personnel as "business" or "companies".

technology actually being used or applied for the first time. *Innovation processes* can be defined as "the invention and implementation of new ideas, which are developed by people, who engage in transactions with others over time within an institutional context, and who judge the outcomes of their efforts and act accordingly" (Van de Ven and Poole 1990). In addition, there are several *types of innovations*, such as product, process, marketing and organizational innovations. This paper will focus only on technological innovations, which can include innovative products and processes. A technological innovation requires a previous decision that is easily identifiable as being determinant for its success, such as acquiring or developing products or technologies, buying intellectual property rights or designing innovation policies.

Decision-making models are conceptual frameworks for understanding how decision-makers process information and arrive at conclusions (Harren 1979). They can be defined as a simplified description of a psychological process in which one organizes information, deliberates among alternatives, and makes a commitment to a course of action (Harren 1979). Although there is no model for decision-making with regard to technology, there is a significant number of different dimensions and models of general decision-making in the scientific literature (see among others Scott and Bruce 1995; Hunt et al. 1989; Harren 1979; Swami 2013). Decision-making is a multidisciplinary topic that spans across many disciplines such as philosophy, psychology, management, economics, engineering, and mathematics. It is therefore not surprising that most dimensions of the models are related to the fields or disciplines of the authors conducting the research, such as health (Smith et al. 2008; Murray et al. 2007), education (Galotti et al. 2006; Harren 1979), military (Thunholm 2004; Scott and Bruce 1995), psychology (Curseu and Schruijer 2012; Schwartz et al. 2002; Starcke and Brand 2012), economics (Simon 1959, 1979; Menzel 2013), and business (Jauch and Glueck 1988; Swami 2013; Schoemaker and Russo 1993; Sull and Eisenhardt 2012).

Given the lack of a decision-making model specific to technology, it is convenient to observe similar examples developed to frame decisions. In one example of an organizational study of decision-making models with senior managers, Turpin and Marais (2004) described nine existing approaches to decision-making: (1) rational, (2) bounded rationality, (3) incrementalist view, (4) organizational procedures view, (5) political view, (6) garbage-can model, (7) individual differences perspective, (8) naturalistic decision-making, and (9) multiple perspective approach. Other studies were less abundant in the number of possible ways to decide. For example, in a military-oriented study, Thunholm (2004) reported on five decision-making styles (independent, not mutually exclusive): (1) rational, (2) intuitive, (3) dependent, (4) avoidant, and (5) spontaneous.[3] However, Thunholm supported the view that although the rational and intuitive styles were unproblematic from a theoretical point of view, the intellectual foundations of the other styles were unclear. For example, his spontaneous style "might perhaps be viewed as a kind of high-speed intuitive decision-making style" (Thunholm 2004, p. 934). Furthermore, the literature reveals that most decision-making models propose, at least, rational-analytical reasoning and an emotional-intuitive approach to explain the major considerations dominating the attention of an individual during this process. Therefore, given the variety of approaches that can be considered in a decision-making model, there is a strong case for using rational-analytical reasoning and an emotional-intuitive approach.

Several authors have argued, furthermore, about the existence of a third political-behavioral approach in the model in order to account for the impact of different stakeholders[4] on the decision-making process (see, for example, Linn et al. 2013; Jauch and Glueck 1988; Ilori and Irefin 1997; Dill 1975; Gray and Ariss 1985; Narayanan and Fahey 1982). According to Cray et al. (1991), political processes that are normally hidden are brought to the fore in strategic decisions that carry high stakes for those involved, affect the organizations in which decisions are taken, and/or might have potential effects on large segments of society. To these authors, the political aspect of decision-making is very important because a "bad" decision can be costly. In fact, an erroneous decision can cost a manager, a researcher or a politician his/her credibility, promotion, bonuses or even her/his job; backing a wrong alternative can cost a department or political faction its political future; and a serious error can accelerate the death of an organization, a department, and even a political faction.

This paper attempts to find a decision-making model entailing three main requirements related to innovation. First, the model should be adequate to where most innovations occur: The business environment and the public research organizational setting. It therefore has to be coherent with the ways business innovators make decisions as well as public researchers. Second, innovation processes need to be analyzed within a model that is compatible at all levels to the interactions between

---

[3]The Swedish General Decision-making Style (GDMS) inventory was created on the basis of work by (Scott and Bruce 1995) and validated with 1441 male military officers with regard to career decision-making and, later, with samples of students, engineers, and technicians with regard to important decisions in general.

[4]The term "stakeholders" is here referred to sensu lato as defined by Freeman and Reed (1983), which includes shareowners, employees, customers, suppliers, lenders, and society.

the three main actors, namely researchers, business innovators, and in particular policy-makers. The model should also be able to accommodate the way interactions between these actors occur at different levels: In the larger context of socio-technical transitions, at the policy level where governments and policy-makers play a decisive role, and at the business level where individuals and organizations are determinant to innovation. Third and last, the model has to be able to encompass the strategic dimension of innovation to researchers (or inventors), firms, governments, and society at large. According to Nutt and Wilson (2010), the term *strategic decision-making* is often used to indicate important or key decisions made in organizations of all types. In this work, the emphasis is on studying the making of a significant technology decision in the context of an innovation. Decisions regarding techno-logical innovation are directly associated with new innovations, which emphasizes the strategic nature of these decisions in a firm, a research lab, or an innovation policy.

Taking these aspects into consideration, it can be concluded that a possible suggestion of a model for a technology decision would need to be based on strategic decision-making and include not only a rational and emotional approach, but also a political one. In their strategic management studies, Jauch and Glueck (1988) identified three major decision-making approaches as being rational, emotional and political. However, my literature review suggests adopting more specific labels for the styles that capture these types of decisions: (1) rational-analytical, (2) emotional-intuitive, and (3) political-behavioral. Figure 1 captures the components of three possible strategic decision-making processes.

- The *rational-analytical approach* (see 1 in Fig. 1) to decision-making is based on the use of quantitative methods. The decision is the choice the actor makes to maximize advantages in full awareness of all available and feasible alter-natives (Jauch and Glueck 1988). In complex cases, it requires close collabo-ration between the analysts and other potential users of the decision. It prescribes a rational, conscious, systematic, and analytical approach.
  According to Dean and Sharfman (1993), rationality is the extent to which the decision-making process reflects a desire to make the best decision possible under the circumstances. According to these authors, this intended rationality (or procedural rationality) is characterized by an attempt to collect the infor-mation necessary to form expectations about various alternatives and by the use of this information in the final decision. Therefore, rationality in strategic decision-making reflects the degree of involvement in collecting information relevant to the decision and of the reliance of the decision-makers on analysis of this information. Furthermore, Kuhlmann et al. (2010) support the view that

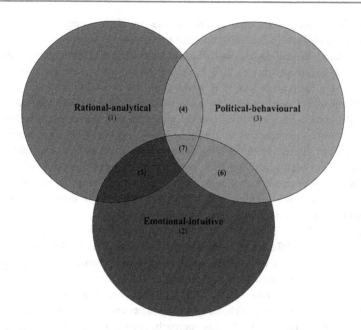

Fig. 1 Components of strategic decision-making processes (adaptation from Jauch and Glueck 1988, p. 23)

intended rationality will remain an innovation actor's prevailing mode of guidance for learning and changing perspectives. Indicators are instruments that can reveal the rational-analytical thinking dominating the individual's attention during the decision-making process.

Jauch and Glueck (1988) pointed out that there are three major criticisms of this theoretical approach: (a) the decision-maker is not alone and is often a part of a multiparty decision-making situation; (b) decision-makers are not rational enough or informed well enough (e.g., gathering information can be too costly to consider for all the alternatives and consequences of the decision); and (c) decision-makers have more goals than just the maximization of objectives. For example, they can be simply aiming to justify their decision instead of finding the optimal solution. It should also be stressed that the objectives might change, which may also undermine the optimal solution.

- The *emotional-intuitive approach* (see 2 in Fig. 1) in decision-making is based on habit or experience, gut feeling and instinct, and is guided by unconscious mental processes (Jauch and Glueck 1988). Many judgements are made by emotional-intuitive decision-makers who are considering a number of alternatives and options. In fact, most decision-makers prefer judgements, as they are considered to lead to a better decision than analytical techniques. This is because judgments can include factors or timing which are not taken into account in the rational-analytical model (Ilori and Irefin 1997).

This model has been criticized because (a) it does not effectively use all tools available to decide, and (b) a rational model ensures that proper attention is given to the consequences of decisions before significant mistakes are made (Jauch and Glueck 1988).[5] Furthermore, (c) there is a lack of scientific consensus as to what exactly intuition means,[6] which partially explains why I opt to name this approach "emotional-intuitive" rather than adopt the authors' term "intuitive-emotional". Other authors just call this approach "emotional", which excludes an enriching part of the concept of intuition.[7]

- The *political-behavioral approach* (see 3 in Fig. 1) in decision-making highlights the pressure of different stakeholders and their impact on the decision-making process. Such stakeholders can be suppliers, trade unions, owners, workers, government, other institutions (parliaments, committees, agencies, etc.), competitors, colleagues, experts, etc. The political-behavioral approach implies that there is a limited number of choices available, determined by the organization and institutional arrangements. In this approach, decisions are made when several

---

[5]In this emotional-intuitive model, the use of indicators is rather limited because it is normally based on qualitative methods.

[6]Some researchers consider the scientific study of intuition impossible, seeing it as an esoteric phenomenon or just erratic nonsense. However, Schoemaker and Russo (1993) considered that intuition can be brilliant when based on extensive learning from past experience, probably reflecting an automated expertise. In fact, the current technical conception of intuition implies that it arises from knowledge and experience. It also implies that intuition involves a form of information processing that might be more implicit than explicit, but which is not at all irrational.

[7]According to Policastro (1999), intuition may be defined as a tacit form of knowledge that guides decision-making in a promising direction, which in the context of innovation leads to potentially creative results. Intuition is assumed to be especially important in tasks with high complexity, short time horizons, ill-structured problems, and involving moral evaluations (Linn et al. 2013). It involves the ability to quickly synthesize and integrate information and use of decision-makers' experience. To Policastro, intuition seems to be most useful when there are high stakes, a high level of uncertainty, and pressure to make the right decision in a limited amount of time. In her perspective, intuitions are not infallible, since they are like rough estimates, which necessarily entail some margin of error. In addition, research showed that there is not much knowledge about how intuition works, under which circumstances it may or may not be useful, or how to reduce its margin of error.

stakeholders agree that a solution has been found through adjustments and negotiations. This approach considers that the real decision-makers must consider a variety of pressures from people affected by their decisions and reflect whether the decision can be implemented politically (Narayanan and Fahey 1982; Ilori and Irefin 1997; Jauch and Glueck 1988). In fact, an organization has to interact with a variety of stakeholders with different amounts of power who give something to an individual or an organization and expect something in return. Naturally, the more power stakeholders have, the more influence they will have on decisions because organizations are more dependent on them (Jauch and Glueck 1988). Therefore, decision-makers meet stakeholders' demands through political compromise, by mutual adjustments, and by merging competing demands to create a coalition of interests that will support the decision.

Decisions are made in a process of mutual influence that may involve actors with different, sometimes even opposing interests. In fact, strategic decisions are often made in a social process of interaction between different actors and groups of actors. In this train of thought, some authors have stressed that, in reality, when decisions are made it is the product of social relations that matters, such as networking activities, different forms of social pressure, and the expression of values and norms. According to Six[8] (2002), decision-makers use only a certain amount of knowledge when making a decision, and their judgments are rather dependent on where actors are socially situated and integrated. Six (2000) supports the view that forces of social regulation and social integration exist and shape the individuals' decisions. These forces can be used to explain how several social actors use information, behave and judge. The product of social relations acts upon individuals, both consciously and unconsciously, constraining and guiding them throughout their decision-making. Therefore, the term 'social relations' is used broadly here to refer to the multiple ways people are connected and influence one another (Hall and Lamont 2013). The term relates to all the social activities that can influence a decision about technology such as networking activities, hierarchical or peer pressure, marketing activities, and values and norms.

This approach has been criticized for implying that institutions play a real role in limiting the choices available to a decision-maker (Jauch and Glueck 1988). It implies that decision-makers accept and recognize the power of stakeholders. However, decision-makers might pretend to negotiate and not accept mutual

---

[8]Editors' note: The author calls himself "6", but "Six" is used here in order to avoid confusion.

adjustments and real negotiations. Furthermore, the approach unrealistically implies that all decision-makers have previously considered whether the decision's outcome might be implemented politically.

Figure 1 illustrates the possible interactions between the three different approaches since the three processes can occur in parallel. The rational-analytical and political-behavioral approaches can interact (see 4 in Fig. 1). In fact, a decision-making process can be simultaneously high or low in politics and high or low in rationality because in some cases it can be rational to behave politically or it can be political to behave rationally (Linn et al. 2013; Dean and Sharfman 1993). Furthermore, politics may frequently obstruct the flow of information, particularly in high-velocity environments where timely and accurate information is only shared amongst selected members of the group (Eisenhardt and Bourgeois 1988; Six 2002). Therefore, political behavior may hinder the rational assessment of a situation since decision-makers cannot consider the whole range of different alternatives (Linn et al. 2013). It is important to stress, however, that to the best of our available knowledge, the scientific literature does not provide sufficient empirical evidence regarding the other interrelations that can exist, as shown in Fig. 1. First, the interaction (see 5 in Fig. 1) between the rational-analytical and emotional-intuitive approaches is based on assumptions by researchers in the literature who failed to provide rigorous empirical data, according to a review made by Linn et al. (2013). The interaction (see 6 in Fig. 1) between political behavior and emotional-intuitive approaches is not known, and according to Linn et al. there are no empirical studies about this interrelation. Finally, Linn et al. (2013) found that a process dominated by rationality and supported by intuition yield more effective political processes (see 7 in Fig. 1). However, the authors failed to back up their findings convincingly, which renders the interaction between the three approaches unconfirmed.

In sum, an analysis of the decision-making literature reveals that most models of decision-making propose at least rational-analytical reasoning and an emotional-intuitive approach to explain the main considerations dominating the attention of an individual during the decision-making process. The proposed model for decision-making on technology also includes a political-behavioral approach, mostly because these decisions can also be understood as strategic decisions in innovation processes. The decision to use a predominant approach can be supported by an inquiry into the influence of indicators. With the exception of the interaction between the rational-analytical and political-behavior approaches, the remaining interactions between approaches to decision-making are not well understood.

2.2 The Rise of Indicators

An indicator can be defined as a numerical sign that shows what something is like or how a situation is changing. An indicator, as a specific sign of a phenomenon under study, is commonly, though disputably,[9] defined within the boundaries of classic scientific disciplines and assumed to be a scientifically objective measure beyond debate and a proxy for scientific knowledge. This paper deals with indicators as quantified instruments available to help people decide about technology in the context of innovation, such as costs, technical characteristics, market share, R&D expenditures, carbon emissions, and size of vehicle fleets.

Indicators are an expression of the human effort to simplify the governance of reality. They are inherently connected with the social need for quantification both for public and for scientific purposes. Porter (1995, p. 74) argued that these efforts at quantification were generally allied with an increase in the "spirit of rigor". According to the author:

> Strict quantification, through measurement, counting, and calculation, is among the most credible strategies for rendering nature or society objective. It has enjoyed widespread and growing authority in Europe and America for about two centuries. In natural science its reign began still earlier. It has also been strenuously opposed. This ideal of objectivity is a political as well as a scientific one. Objectivity means the rule of law, not of men. It implies the subordination of personal interests and prejudices to public standards (Porter 1995, p. 75).

The origins of indicators as modes of knowledge and governance can be traced back to the practices of business management at least since the fourteenth century, where quantification was crucial to the trade Venice conducted with Cyprus (Maddison 2003). Later, the rise of the scientific mentality played an important role, since it included an insistence on objectivity, on the written word, on rigorous quantification, on transparency, etc. (Porter 1995). Furthermore, the creation of indicators is also associated with the need for governance, which originated in the seventeenth century when "the art of reasoning by figures on things relating to government" was called "political arithmetic" (Maddison 2001). Later, indicators

---

[9]In fact, although commonly accepted, debates continue about the object measured by indicators (such as non-R&D innovation expenditure, SMEs introducing marketing/organizational innovations, and innovative SMEs collaborating with others), as well as the nature of reality being measured. Furthermore, an indicator's claim to objectivity has to be restrained to "knowledge produced in conformity with the prevailing standards of scientific practice as determined by the current judgements of the scientific community" (Porter 1995, p. 216).

were used in connection with the rise of the modern nation-states in the beginning of the nineteenth century, and their need to govern objectively, impartially, and transparently (Merry 2011). Most of the initial demands for indicators came from engineers and technocrats who were committed to the development of an administrative culture in modern France, highly interested in management, and with considerable enthusiasm for, among others, Frederick Winslow Taylor's work on efficiency (Porter 1995).

In the 1930s, a significant impulse to quantification came from the growth of science in the US and the culture of equidistance and impersonal objectivity in decision-making. In America, important efforts at quantification were made with the systematic use of Intelligence Quotient[10] tests to classify students, opinion polls to quantify the public mood, elaboration of statistical methodologies for licensing drugs, and even cost-benefit and risk analyses to assess public works (Porter 1995). Later in the 1950s, the need for objectivity and quantification was driven by the US Corps of Engineers. The Corps began concentrating the economic efforts of their "failed engineers" on all district offices "where they were likely to do less harm" (Porter 1995, p. 186). At the same time, the Corps began employing increasing numbers of economists and other social scientists, which led to a takeover by economists and the emergence of cost-benefit analysis in modern economic studies. This type of analysis began with water projects and transport studies, but was later disseminated particularly through RAND[11] military studies. Power (1997) and Porter (1995) suggest that accountants and the rise of bureaucracy also played a significant role in the rise of quantification in most aspects of life, particularly during the twentieth century.

In the 1970s, a "social movement of indicators" was born in reaction to the economic orientation of the above-mentioned quantification culture. This group of intellectuals, mostly social scientists, considered that the word "social" was restrictively defined and meant only "outside the realm of economics" (Sheldon and Parke 1975, p. 695). To them, science (or quantification) created a new sort of

---

[10]Intelligence quotient, commonly known as IQ, is a score derived from standardized tests designed to assess intelligence.

[11]According to Linstone (2008), the Rand Corporation is the most influential American think tank of the second half of the twentieth century. The company was established in 1946 to deal with useful applications labelled "operations research", which applied mathematics to problems such as interceptor vectoring and convoy protection. Presently, Rand is well accepted in decision-making corridors of Washington, DC, and offers vast research and analysis to the US armed forces. Rand is currently financed by the US government, private foundations, corporations (including the health care industry), universities, and private individuals.

"philistines" encouraged by the relative ease of expressing quantities in dollars and, consequently, given an exaggerated importance in the interpretation of the realm. Therefore, these scientists argued for a change in conceptual frameworks, shifting the emphasis from economics to measures of social change, which included different subsystems in society like education, health, and economics. The arguments were centered on the controversy that resulted from an economic-centered perspective, and mainly focused on what this new wave of indicators should measure.

The controversy eventually faded away, and the quantification rationale started to be applied to all parts of life, broadening its scope to almost every area of knowledge or human endeavor. Today, we can find this quantification rationale applied on a global scale (Misuraca et al. 2013; Maddison 2003) and even in our personal lives. For example, Robichaud et al. (2006) reported on the existence of indicators for one's personal quality of life, such as measures of verbal communication interaction, well-being, level of participation in social activities or involvement in the community and integration in the community. This new quantification trend should not be interpreted as a conspiracy against the culture of qualitative reasoning, as some in social science circles have argued. In fact, the production of an indicator reflects values that existed in society before the creation of their moral, ethical, political, economic, or financial nature. In fact, their significant expansion in recent years was based on an existent societal will, expressed for example by many influential organizations (e.g., OECD, ILO, UNDP, WHO, universities, etc.). Today, even a qualitative analysis of a large research question has to include an attempt to quantify aspects of an issue as much as possible. In fact, the present work aims to present quantified data on the influence of indicators in specific decisions in order to complement further qualitative analysis.

2.3 Indicators in Decisions on Technological Innovation

As mentioned earlier, indicators have been employed in societal debates for a long time, embedded in a culture of objectivity. They are considered to be a useful instrument to guide and improve our collective and personal decisions. While indicators are ostensibly present in our collective rhetoric and apparently affect our day-to-day lives, comparatively few studies have been carried out addressing the relevance of indicators in management, research and policy, particularly in the context of innovation processes.

The extent to which indicators are used in decision-making is largely unknown. Most of the literature aims to develop indicators, to analyze them, or to evaluate them. Only a few authors have provided clues about the extent to which they are used to make a decision, and these qualitative studies were mostly restricted to policy-making. In fact, Gudmundsson and Sørensen (2012) found evidence that indicators seem to play a very limited role in policy-making about sustainable transport. Furthermore, MacRae (1985) argued that the most frequent problem of indicators is their non-use in policy-making. The reasons for this disregard of indicators can be found in the lack of interest, in information overload, in a lack of communication, or even in opposition to what is being measured. It seems, therefore, necessary to proceed in two directions: First, broaden the investigations to other areas of decision-making to allow for comparisons with policy-making, and second, determine the role of indicators in decision-making because no quantified information exists to provide an initial background for analysis. The study of indicators in decision-making regarding technological innovation will help characterize the social phenomena behind the process of technological innovation. It will provide information about the main actors, the use of indicators and their real influence, the role of the social context, and the role the type of decision plays in the decision-making process.

2.3.1   Use and Influence in Decision-Making

It is necessary to disentangle the use of indicators from their factual relevance in decision-making, as suggested in the Gudmundsson and Sørensen (2012) study. In fact, determining the real influence of indicators makes it not only possible to understand their importance in the decision-making process, but also to determine which approach used in making the decision was dominant. In turn, the investigation will provide clues about how knowledge is introduced in decision-making and a deeper understanding of the social process of decision-making in the context of technological innovation.

Unfortunately, questions about the effective influence of indicators have seldom been the object of research. The seminal work of Gudmundsson and Sørensen (2012) revealed that the *use* of indicators does not automatically mean that they exerted an *influence* on policies or processes with regard to the example of sustainable transport policy. In their study, indicators played a very limited direct role in the decision-making processes. Generalization is difficult, however, particularly because sectorial conditions might change significantly from transport policy to, for example, pharmaceutical policy, where a significant amount of quantified data and objective information exists and is accessible. Therefore, it is also necessary to broaden the scope of these sectorial and policy-related findings to include other decision-making processes.

This paper intends to contribute findings providing evidences for the use and influence of indicators in the decision-making process of knowledge-intensive innovation groups. It will be assumed that indicators are influential in decisions on technological innovations made by knowledge-intensive groups because rationality-driven decisions are expected to predominate among those who participate in technological innovation without any noteworthy knowledge constraints.

The influence of indicators in the context of policy-making was categorized by Gudmundsson and Sørensen (2012) in four groups: Instrumental, conceptual, process, and symbolic. The authors considered, first, that an instrumental role of an indicator means that it had a direct influence and was used directly as a tool to form a decision. Second, a conceptual role reveals that indicators contributed to shape knowledge or introduce new ideas, but were neither used directly nor influential in decisions. Third, a process role means that an indicator used over time affects the way some aspect of policy-making is conducted, regardless of what the indicator tells/reveals directly. Fourth and last, a symbolic role means that an indicator was used to justify decisions that have already been taken or to give the decisions the appearance of rationality. According to the authors, this role may still embody an influence on policy in terms of policy legitimacy or in terms of the balance of power.

However, these second and third categories present significant difficulties, probably because they were formulated to analyze a specific policy-making case and not to examine decisions regarding technological innovations. There are four main difficulties. First, there is a significant difficulty in distinguishing a conceptual from a process role in two ways: (a) an indicator that contributes to shaping knowledge or introducing new ideas (conceptual role) may also, in some aspects, affect the way of how policy-making is conducted (process role); (b) an indicator may be used over time (process role) and may neither be used directly nor be influential in a decision (conceptual role). Second, a conceptual role probably does not exist in the practice of a technology decision-maker. For this individual, an indicator that contributes to shaping new knowledge and ideas inevitably changes the framework of a decision—for example, through the previously mentioned constitutive effects—, and consequently exerts influence on the decision. In other words, for a scientist, an engineer, a manager, or a policy-maker reflecting on the influence of an indicator in a technology decision, it is difficult to say that an indicator was not influential if one admits that it has contributed to shaping knowledge and introducing new ideas. Third, it is difficult to admit that an indicator influenced some aspect of the decision regardless of what it evidently means. Fourth and last, from a practical point of view, an indicator is an objective concept that is either present or not in a concrete technology decision. An indicator can hardly be a semi-part of a decision, such as by shaping knowledge or introducing new ideas used over time without being immediately used in a decision. Bearing these four difficulties in mind, it can be concluded that these two distinctions are not of substantial interest with regard to decisions about technological innovations.

In this paper, I will only use the instrumental and symbolic roles to understand the influence indicators have on technology decisions. Indicators can thus either play (1) an instrumental role, (2) a symbolic role, and (3) no role at all in decision-making about technological innovations:

An *instrumental role* (1) means that an indicator exerted a direct influence and was used as a tool to make a technology decision. The instrumental influence of an indicator in a decision implies that a rational-analytical method was used, such as evidence-based analysis or operational research. In addition, the indicator's role also suggests that a rational-analytical approach was employed exclusively or was predominantly complemented by a political-behavioral or emotional-intuitive approach.

A *symbolic role* (2) means that indicators were used to justify decisions after they had already been taken or to give decisions the appearance of rationality. The symbolic role implies that no rational-analytical method was used predominantly in making a decision. This role indicates that the predominant decision approach was political-behavioral, either exclusively or predominantly in interaction with a rational-analytical or an emotional-intuitive approach. In this train of thought, it can be argued that the desire for quantification is mostly symbolic at the policy level. In fact, Six (2002) claimed that decision-makers use a certain amount of knowledge when making a decision, and their judgments are dependent on where the actors are socially situated and integrated. This view implies that the influence of indicators is determined by the social "cage" of the users and that social relations act upon individuals, constraining and guiding them both consciously and unconsciously in making decisions. However, the influence of these social relations can vary in different social groups, particularly in groups where the social cage can be less present, such as the researchers group. Therefore, to understand the role of indicators it is necessary to inquire about the role of social relations in decision-making.

Last, situations in which indicators plaid *no role* in the decision-making (3) were associated with conditions where information was lacking, incomplete or ambiguous and thus could not be analyzed, or where the consequences of the decisions were significantly unpredictable. Such conditions are associated with predominantly emotional-intuitive approaches, such as judgements or gut feelings

(Jauch and Glueck 1988). The absence of indicators also reveals a lack of rational-analytical tools[12] or quantified measures in the decision-making process. Therefore, the case where indicators play *no role* shows that there is an emotional-intuitive approach, whether exclusively or in dominant interaction with a political-behavioral approach.

It is important to take into consideration that the indicators used in technology decisions can be very different. The major difference can be found at the level at which they are used in the innovation processes. First, policy-makers make technology decisions that influence the innovation system mostly at the national level. For example, policy-makers can use the indicator business enterprise expenditure on R&D (million current PPP $[13]) to decide upon a policy to stimulate university-company R&D projects. Second, business R&D&I personnel make technology decisions of a different sort. The type of indicators this group tends to use is mostly (but not only) related to the market or company because these technology decisions tend to be connected to the development of new products or technologies. For example, such a decision can use indicators related to innovation in a firm, such as the number of new products using their patented technology, or the costs of hiring a new researcher to develop a new technology. Third, (public) researchers, academics, and R&D health personnel tend to make decisions mostly at the product or technology level. For example, in a R&D project a researcher will focus on an indicator of the financial cost of a new DNA sequencer[14] or microscope to further her/his research. Therefore, the type of indicators used when making technology decisions during innovation processes can vary significantly and is primarily related to the type of decision.

Within this diversity, however, it can be useful to link an indicators' use to a technology decision for two reasons. First, an indicators' use, use in combination with other factors, or non-use can indicate the real influence of indicators in decision-making about technological innovations and uncover other drivers or motivations such as the pressure from other groups. Second, the theoretical model

---

[12](i) Rules of thumb—for example, if the rule is: If there is a 20 % increase in profits we need to buy new technology; (ii) evidence-based analysis—for example, when making cost-benefit analyses, weighting options, etc.; and (iii) operations research—using mathematical models to explore quantified evidences. All three methods require some sort of quantification, which would mean having an indicator of something.

[13]Purchasing Power Parities (national currency per dollar).

[14]A deoxyribonucleic acid (DNA) sequencer is a scientific instrument used to automate the DNA sequencing process. A sequencer is used to automatically determine the order of the DNA's four constituents: Adenine, guanine, cytosine, and thymine.

previously suggested for technology decisions in the context of innovation can be tested by identifying the roles of indicators. Determining that indicators played an influential role in a decision indicates that a rational-analytical method dominated the individual's attention during the decision-making. A symbolic role of indicators (e.g., when social relations were considered more important than indicators) reveals that a political-behavioral method dominated the decision. And an emotional approach was used when indicators were not used at all and social relations were also not significantly relevant.

2.3.2 Types of Technological Decisions in the Context of Innovation

There are four types of technology decisions that can occur in innovation processes. In fact, in the context of innovation there can be, first, decisions about the acquisition of new equipment and/or technology. Second, innovations can also occur after decisions have been made about the development of a new product and/or a technology. Third, innovations also can be the outcome of decisions about the acquisition of intellectual property, which can involve buying patents, trademarks, industrial designs, geographical indications, or copyrights. Fourth and last, there can be decisions about the design of policies which can produce effects in the innovation system and, consequently, in technological innovation.

Any decision is expected to contain elements of complexity and uncertainty. There are several reasons that these two factors are central to technological innovation. First, technological innovation is often associated with complexity[15] (Chapman and Hyland 2004; Rycroft 2007; Waelbroeck 2003; Wonglimpiyarat 2005). In the context of technological innovation, an example of complexity is when components that are integrated together cause difficulties in the transformation into successful products/processes (Wonglimpiyarat 2005). Complexity in innovation has been associated with experiences where information is incomplete, ambiguous, and the consequences of actions are highly unpredictable (Aram and Noble 1999). In these contexts, complexity is contained in the technologies, products, customer interfaces, and organizational setups (Chapman and Hyland 2004). Second, technological innovation is also often associated with uncertainty (Dosi 1982; Nelson and Winter 1977; Carbonell and Rodríguez-Escudero 2009; Sainio et al. 2012).[16] Uncertainty can be defined as the degree to which a number

---

[15]According to the Oxford Dictionaries Online (2014), complexity means the state or quality of being intricate or complicated.

[16]See also Böhle (2011), Jalonen and Lehtonen (2011), Fusari and Reati (2013), and Meijer et al. (2007).

of alternatives are perceived as uncertain with respect to the occurrence of an event and the relative probabilities of these alternatives (Rogers 2003). Innovation involves uncertainty in an essential way because "different people, and different organizations, will disagree as to where to place their R&D chips, and on when to make their bets" (Nelson and Winter 1977, p. 47). There are numerous types of uncertainty associated to innovation processes although technological, market and regulatory uncertainties have an established status (Jalonen and Lehtonen 2011; Sainio et al. 2012).[17] It can be concluded that uncertainty is a central element in decisions regarding technological innovation.

Each type of decision presents different levels of complexity and uncertainty. First, decisions about the acquisition of equipment or technology are less complex than the three following types of decisions because they primarily involve comparisons between products or technologies and immediate assessments of the impact in a technological innovation process. Furthermore, the level of uncertainty associated with this type of decision is also rather low since most buyers know what to expect from the acquisition of equipment or technology. Second, decisions about the development of a product or technology are complex when compared to the other types of decisions regarding technological innovation (Chapman and Hyland 2004). They are often a matter of complex strategy and require not only a decision about what product or technology should be developed and why, but also knowledge on how to pursue with the development phase of a product/technology. Furthermore, this type of decision frequently involves a significant level of uncertainty, mostly because there is no guarantee of success (Böhle 2011). In fact, the development of a product or technology may be subject to changing circumstances that might render a technology less attractive or even obsolete, and where technological promise may even never materialize (Kemp et al. 1998). Third, decisions about the acquisition of intellectual property rights are simpler and less uncertain than the other three types of technology decisions. Although they may involve some strategic thinking, they are often based on the acquisition of knowledge about a workable solution that an industrial patent offers to a problem or a need for a product or a technology. Furthermore, the acquisition of propriety rights often carries less uncertainty than the development of a product or technology. In fact, the technological conception has already been proven and accepted

---

[17]But many more can be identified. For example, Carbonell and Rodríguez-Escudero (2009) considered only two aspects of uncertainty: Technology novelty and technological turbulence. In their study of innovation on biomass gasification projects in the Netherlands, Meijer et al. (2007) reported that technological, political and resource uncertainty are the most dominant sources of perceived uncertainty influencing entrepreneurial decision- making.

by patent offices, even though the decision still carries the risk that a technology may become less attractive or obsolete and that the expectations may not materialize (Kemp et al. 1998), for example. Fourth and last, decisions about policy design are very complex and uncertain because they involve concerns about the social and economic impact of a complex and uncertain target, namely the innovation system (see Smits et al. 2010; Kuhlmann et al. 2010).

In summary, there are four types of technology decisions that can occur in innovation processes: Acquisition of new equipment and/or technology, development of a new product and/or technology, acquisition of intellectual property, and policy design. Any decision is expected to contain different elements of complexity and uncertainty. The use of indicators is negatively associated with complexity and uncertainty. Therefore, different types of decisions require different indicators.

To conclude, decision-making processes are complex phenomena where there may be several factors exerting influence. Decision-making processes depend not only on the contextual factors but also on factors associated with the type of decision. A quantitative study of the role of indicators in decision-making about technological innovation can help bridge the gap in our knowledge about indicators in the context of decision-making. This study aims to distinguish between the use and the real influence that indicators actually have on decisions, detect other influences, and identify the types of technology involved in the decision.

3   Methodology

This research focuses on the initial technology decision which is made after an idea is born and the preliminary analysis of its benefits is carried out. The focus on the initial decision can help to understand how decisions are made in technological innovation processes. The term technology decision here is related to a decision made in the initial phase of an innovation process, such as (1) a decision to acquire equipment or a specific technology, (2) a decision to develop a product or a specific technology, (3) a decision to buy property rights, or (4) a decision related to the design of technology policies (programs, measures, actions, projects, etc.). These types of decisions can occur across all sectors, company or research center sizes, levels in organizations, different cultures, etc.

The research undertaken in this study targets the three groups in an innovation system:

1. Researchers: A group composed of public researchers, academics, and R&D health personnel, hereafter named *researchers*.

2. Business research, development and innovation (R&D&I) personnel: A group composed of team leaders of R&D&I departments in companies, hereafter named *business* or *companies*.
3. Policy-makers in the field of technological innovation: A group involved in technology decisions and the design of strategies that address the innovation system (Kuhlmann et al. 2010).

There are both general and specific reasons to target these three groups. Common reasons for selecting these groups were: First, the groups are composed of individuals who (i) reported on recent technology decisions (e.g., acquisition of equipment, development of products or technology, investment in intellectual property, or involvement in the design of innovation-related policies); (ii) have access to and use vast amounts of techno-scientific knowledge; (iii) normally have contact with indicators; and (iv) possess the skills to deal with indicators. Second, the way these individuals use indicators captures the extent to which knowledge of facts was examined in order to come to a decision. It also reveals to what extent this type of knowledge is applied in practical terms in the decision-making process. Therefore, research on the use of indicators in the decisions can contribute to an understanding of decision-making and knowledge management in innovation contexts. Third, these groups are closely linked to the innovation system and play important roles in its functioning. All of them deal with the production, management or dissemination of innovation. It is expected that this line of research will contribute to our understanding of the interactions between technological innovation and society.

Furthermore, the three groups were individually targeted because:

1. *Researchers* can influence technology developments in different ways. First, they are involved in R&D projects that can lead to new technologies and innovations. Second, the group is involved with students before they reach decisive positions and influence other researchers and other communities with regard to technology options. Third, these individuals are frequently called to decide upon significant investments that have a direct or indirect influence on technology options, such as research programs and projects, setting up new laboratories, universities, hospitals, clinics and other specialized institutes, acquiring powerful microscopes and diagnostic machines, etc. Fourth and last, this group informs and helps decision-makers regarding technology choices. For example, they can assist in important public and private choices, such as the development of satellites, transport systems, other research-related facilities,

the acquisition of submarines, supercomputers, helicopters and airplanes, the construction of highways, bridges and dams, etc.

2. *Business R&D&I leaders* are responsible for most choices regarding the development of new products or new technologies, and are frequently in charge of important strategic decisions inside the most innovative companies. In fact, they are most frequently found leading innovation departments of firms or in charge of R&D projects.

3. *Policy-makers* in the field of technological innovation are involved in technology decisions and the design of strategies that address the innovation system (Kuhlmann et al. 2010). Their policies can involve programs, projects, ideas, legislation, and other regulatory frameworks that affect the development of technological innovations. These policy-makers deal with policies and legislation designed, for example, to promote science and technology graduates, support patenting efforts, backing export-oriented companies, etc.

Various methodologies were applied to ascertain the influence of indicators in technological innovation. As mentioned before, the initial work combined literature research, analysis of official documents, and (a) exploratory interviews. Afterwards, the work included (b) the design and administration of three closed online questionnaires, which is the focus of the present work. The questionnaires were complemented with (c) seventeen semi-structured interviews. In addition, the work included (d) eight in-depth interviews with experts conducted after the surveys, yet in this paper the information from the interviews will only be used to contextualize quantified data. In greater detail:

(a) The exploratory interviews were designed to prepare and calibrate the questionnaires and subsequent interviews. Twelve experts with sound knowledge and experience on the use of indicators and decision-making were interviewed in Germany, Switzerland, and Portugal between October 2011 and January 2012.

(b) The questionnaires addressed Portuguese representatives of the three groups under analysis, collecting ninety-one valid answers from February 2012 to June 2013. The samples were composed only by individuals who were involved in consistent technology decisions and belonged to the mentioned groups. The samples were created using non-probability sampling methods (Saumure and Given 2008) in the following way:

(i) The samples from the business R&D&I and the researchers' groups were selected using a purposive criterion sampling method (see Palys 2008) based on the 2010 National R&D Survey.[18] The criteria used required the existence of (1) scientific leaders of an R&D project in 2010 with significant expenditures for equipment (i.e., an expenditure with instruments, equipment or software equal or higher than 3000 € for researchers and 1500 € for the business R&D&I personnel group), and to have (2) at least one PhD in the research team (full-time equivalent). In this way, the criteria confirmed not only that a consistent technology decision was made in the context of a concrete innovation, but also that the research team had the proper skills to conduct a sound R&D project. Therefore, the samples were significantly representative because they were only composed by scientists (or equivalents) who took consistent technology decisions in a reliable R&D environment. In this context, there were 57 leaders of R&D&I departments in companies in the National R&D Survey database, and 36 responses were received corresponding to a response rate of 63 %.[19] There were 78 researchers that met the criteria, and 31 responses were received corresponding to a response rate of 40 %.

(ii) The sample of policy-makers was created using snowball sampling (Morgan 2008), because there were no other sources to locate the members of this population and almost all members knew each other. The initial set of research participants (serving as informants about potential participants) was significantly diverse and was complemented by Google searches to avoid any possible bias. Two successive waves of snowballing assured a more representative sample (Schutt 2008). The final sample was composed by the vast majority of individuals directly involved in innovation policy decisions in the period 2005–2011 in the country. The sample can be considered significantly representative, particularly given the reduced number of individuals involved in innovation policy at the

---

[18]The National R&D Survey (named IPCTN) is a reliable long-term survey that captures in detail data about any existing R&D projects and about researchers and companies involved in R&D in Portugal. The survey has also internationally comparable standards, is based on the Frascati Manual, and is regularly checked by the OECD, Eurostat, and the National Institute of Statistics.

[19]The 2010 National R&D Survey database detected 59 companies in the country that met the criteria. The business questionnaire was sent, however, only to 57 due to the closure of two firms.

time. In this context, there were 59 individuals listed as being involved in consistent technology decisions regarding innovation policy. Exactly 24 responses were received, which corresponds to a response rate of 41 %.[20] The response rates obtained for the three groups were relatively high compared to normal standards in social research (see, e.g., Shih and Fan 2009 and Baruch 1999, who report comparatively low response rates in social research). Furthermore, to account for effects on response rates (see Sauermann and Roach 2013; Fan and Yan 2010), there were three personalized email campaigns with three reminders each and personal contacts from February 2012 to mid-June 2013 to encourage individuals to answer. Participants were not compelled by law to answer, there were no economic incentives to answer, and the questionnaires were not part of the national statistical system. Only volunteers could reply to the questionnaires given the scientific nature of the surveys. In addition, the answers given online did not allow us to establish any relationship with specific individual technology decisions. Finally, the questionnaires were short (taking on average five minutes to complete), identical (see below), with closed questions (with two exceptions for the policy-makers group, see below), and always related to concrete technology options.

The three questionnaires were similar. The difference between the questionnaires to researchers and business R&D&I group was small and consisted in small adaptations of the text to the context of each innovation group. However, as mentioned before, there were two distinct features in the case of policy-makers. These features were based on the need for them to assist in the assessment of the degree to which policy-makers made use of indicators in making decisions. First, a pre-question was introduced to identify the position occupied during the technology policy decision. The options were minister, secretary of state, political advisor to the minister, political advisor to the secretary of state, consultant, director-general, civil servant, parliamentarian, city mayor, and other position. Second, three questions were added to understand the level of knowledge about the indicators used in the policy decision: (i) "Please name the indicators used or recommended to make the decision"; (ii) "Did you use another type of information in the decision (e.g., studies, advice or sectorial information)?"; and (iii) "In case you used other type of information, please name the indicators that were used". All responses to the questionnaires were collected anonymously, with the exceptions of those

---

[20]The two sampling techniques allowed the identification of 65 policy-makers but, after a significant number of attempts to locate the policy-makers, six of them were considered to be unreachable.

that resulted from direct interviews. No significant differences were detected between online replies and direct answers.

(c) The twenty-six semi-structured interviews were conducted to complement the data in the questionnaires. The majority of the interviews (13 interviews or 50 %) targeted the policy-makers, followed by researchers (seven interviews or 27 %) and the business community (six interviews or 23 %). The emphasis on interviews with policy-makers was justified by the sensitive nature of the information requested and to avoid any suspicion of misuse of information. All interviews to the three groups were conducted using one guiding questionnaire in a confidential way and lasted on average one hour. The use of the questionnaires as a tool to guide the interviews was a practical way to have similar questions across the three groups, to focus the interview on the critical questions, and to allow new questions to arise during the conversation. In this paper, the interviews will be used only to help contextualize quantified data.

(d) Last, the work also included four in-depth interviews to answer remaining questions with experts, conducted in Portugal and Germany after the surveys, i.e., from April 2012 to June 2013.

As mentioned previously, there were no records of quantitative research applied to the use and influence of indicators in the technological innovation literature. Therefore, the research presented here focused on the quantification of the major characteristics of the use and influence of indicators by surveying only individuals directly involved in technological innovation. These efforts allowed us to establish the initial background necessary to understand the reach of indicators (and of rational-analytical decision-making) in the context of technological innovation. Although beyond the scope of this paper, further qualitative research will complement the quantified results presented in the next section.

4 Results

In this section, I present results from a survey designed to understand the role of indicators in decisions regarding technological innovation. The quantification of their use and influence will establish the initial background of information necessary to understand the presence of indicators in decisions about technological innovation. The work presented here constitutes a preliminary quantitative assessment necessary to further the qualitative study of the role of indicators in decisions regarding technological innovation. It will allow a deeper understanding of the extent to which indicators are present among those who can use indicators as freely as possible in technological innovation. It will also help us understand the main type of thinking dominating the attention of these knowledge-intensive innovation groups during decisions.

The following sections present results related to questions on the above-mentioned questionnaires. The first one addresses the use and influence of indicators in decision-making, describing their presence in each group as an instrument in decision-making, and the weight or the real influence that indicators have in the final decision. This distinction between use and influence is relevant because it reveals the presence of indicators in the decision, and it shows their real contribution to the decision. It will subsequently allow the roles of indicators to be categorized and will help to detect the prominence of decision-making approaches. The second section refers to the context of the decision, by identifying the main external sources of influence in these decisions. As the latter question came before the former in the questionnaire, it enabled the participant to first reflect on the individuals who influenced the final decision and afterwards reflect on what was more important: The indicators or the groups previously identified. The third and last section deals with the influence of the types of technology decisions on the use of indicators. As previously mentioned, the types of technology decisions are considered relevant for understanding the need for indicators to be employed in making a decision.

4.1 Use and Influence of Indicators

The first question of the questionnaires was directly related to the use of indicators during a significant technology decision. Table 1 summarizes the answers and percentages obtained for the question.

Table 1 Number and percentage of answers by group to the question "Did you use indicators during technology decisions?"

	Yes		No		Total	
	Answers	%	Answers	%	Answers	%
Researchers	22	71	9	29	31	100
Business R&D&I	32	89	4	11	36	100
Policy-makers	22	92	2	8	24	100
All groups	76	84	15	16	91	100

The table reveals a majority pattern of use of indicators among those who were invited to reflect on the use of indicators during their technology decision-making process. In fact, the vast majority (84 %) of all answers revealed that they used indicators during their technology decisions. Only 16 % of the respondents said that they did not use them. Taken together, the results present a defined pattern, while a closer look into the three groups under observation reveals minor differences worthy of analysis. First, the policy-makers presented the most polarized pattern, in which 96 % answered that they used indicators in their decisions, and only 4 % stated that they did not use them. Second, 89 % of the business R&D&I personnel answered that they used indicators to decide, and 11 % indicated that they did not use them in the technology decision process. Third, the polarization in the group of researchers was less intensive. In fact, a less polarized but still vast majority (71 %) of the researchers answered that they used indicators while making their technology decision. Accordingly, a larger but still minority percentage of answers (29 %) stated that they did not use indicators. Therefore, it can be said that the intensity in the use of indicators in technology decisions is high although different in each group.

In sum, the significant use of indicators detected in the results supports the view that these groups use indicators as an instrument for making technology decisions. These results suggest that this intensive use of indicators translates the quest to use knowledge during the decision-making process. However, the use of indicators should hypothetically mean that they were influential in the decision. This is addressed next.

The next question referred to whether indicators were more important than social relations during the decision regarding technological innovation. Table 2 summarizes the responses by group.

The table reveals that social relations were more important than indicators to the majority (59 %) of respondents. In more detail, the majority of both the policy-makers and the business groups (68 % and 59 %, respectively) answered

Table 2 Number and percentage of answers by group to the question "Do you think that indicators were more influential than social relations during the technology decision?"

	Yes		No		Total	
	Answers	%	Answers	%	Answers	%
Researchers	11	50	11	50	22	100
Business R&D&I	13	41	19	59	32	100
Policy-makers	7	32	15	68	22	100
All groups	31	41	45	59	76	100

that the social relations were more important than the use of indicators. However, researchers' answers were balanced equally between "yes" and "no" (50 %), stating the indicators were as important as social relations in their decision-making process about technology.

Interestingly, a comparison between Tables 1 and 2 reveals that the more indicators are claimed to be used, the less influential they are in decisions concerning technological innovation because (a) policy-makers used more indicators (92 %) but considered indicators less relevant than social relations (68 %); (b) business R&D&I leaders used slightly fewer indicators (89 %) but considered them to be slightly more relevant although less important than social relations (59 %); and finally (c) researchers used fewer indicators (71 %) but considered them as influential as social relations (50 %). Therefore, the disentanglement of use and influence of indicators allowed a concrete identification of the real and different weight indicators carry in the decisions of each group.

The relative significance of indicators among researchers suggests that the social activities of researchers are less relevant to the decision. Several reasons may account for this, such as that the work of researchers tends to be (a) comparatively more autonomous (less dependent on their social context) and (b) more involved in the search for factual objective knowledge to reach scientific conclusions (an activity closely connected with the use of indicators). In contrast, business individuals can be more dependent on company hierarchies and market considerations to make a technological innovation decision; and policy-makers are relatively more restrained by their social activities and are in significant need for expertise (i.e., for indicators as a source of knowledge) in order to be able to decide. The next section will present results to clarify these issues.

These results allow for the categorization of the roles of indicators in decisions, described in Sect. 2. Table 3 presents results related to the role of indicators in technology decisions by groups.

The table reveals that indicators were instrumental to about one-third of the decisions in the groups (i.e., 29–36 %). It also shows that indicators played mostly

Table 3 The influence of indicators in technology decisions by group

	Instrumental		Symbolic		No role		Total	
	Answers	%	Answers	%	Answers	%	Answers	%
Researchers	11	35	11	35	9	29	31	100
Business R&D&I	13	36	19	53	4	11	36	100
Policy-makers	7	29	15	63	2	8	24	100
All groups	31	34	45	49	15	16	91	100

a symbolic role in decisions, particularly in the policy-maker and business R&D&I groups (63 and 53 %, respectively). Results for researchers were distributed more evenly between instrumental and symbolic (each 35 %). Only a few indicated that indicators played no role in the technology decision, although the percentage was higher (29 %) in the case of researchers.

In sum, these results suggest that indicators exerted a symbolic influence mainly among policy-makers as predicted in the literature. Business R&D&I leaders revealed a similar profile, although less emphatically. Researchers used indicators more heterogeneously.

4.2 Influence of the Context

As mentioned previously, the social context is expected to influence the role that indicators play in decisions concerning technological innovation. The intention of the next question was to identify the relevant importance of different individuals or groups to the decision-making. Table 4 presents the results as to the groups exerting the most influence on the decision.[21] This is a question where respondents had to classify the intensity of the influence on a four-point Likert scale to force a choice identifying the most important influence in the decision.

The table reveals a different level of influence for each group. First, researchers indicated that *experts* were their most important influence on their technology decision (52 %), followed by *technology users* and *researchers/academics* (*ex aequo* 42 %). These responses point to the primacy of knowledge sources and users as exerting the most influence on the decisions of this group. Second, the business R&D&I leaders indicated that *management/managers* (69 %) exerted the most important influence, followed by *consumers* (42 %). These answers suggest that hierarchies and users were the main influence in this group. Third, policy-makers indicated that a higher hierarchy level (*other policy makers*) (63 %) and, to a lesser extent, experts (29 %) were the most influential groups (perhaps explaining a need for expertise). These answers suggest that particularly hierarchies and sources of knowledge exerted the greatest influence on this group. Figure 2 summarizes these results, aggregating the main types of influence during technology decisions.

The figure reveals that the groups are primarily influenced by hierarchies, sources of knowledge, and users along three different lines. First, hierarchies were

---

[21]Only results related to "Very important" classification are shown in the table.

Table 4 The individuals/groups exerting the most influence on technology decisions, by group and overall

	Researchers		Business R&D&I		Policy-makers		All groups	
	Answers	%	Answers	%	Answers	%	Answers	%
Higher hierarchy level	7	23	25	69	15	63	47	52
Financial directors and accounters	5	16	4	11	4	17	13	14
Experts	16	52	11	31	7	29	34	37
Colleagues	11	35	9	25	3	13	23	25
Personal relationships (acquaintances, friends, etc.)	4	13	2	6	3	13	9	10
Technology users	13	42	14	39	4	17	31	34
Salesmen/Account managers/Consultants	0	0	3	8	0	0	3	3
Business/Industrial groups	4	13	4	11	4	17	12	13
Researchers/Academics	13	42	4	11	3	13	20	22
Other political decision-makers	0	0	2	6	3	13	5	5
Consumers	4	13	15	42	4	17	23	25
Groups of citizens (associations, pressure groups, etc.)	1	3	1	3	3	13	5	5
Society in general	2	6	0	0	2	8	4	4
Media	1	3	0	0	1	4	2	2
Other	0	0	0	0	0	0	0	0

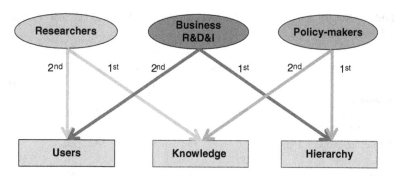

Fig. 2 The groups (*bottom*) exerting the most influence on technology decisions for each group of decision-makers

the most relevant influence in the decisions of business R&D&I leaders and policy-makers. In fact, these two groups declared a higher hierarchy level (*other policy makers, managers*) to be the most relevant influence on their technology decisions (first choices). Second, sources of knowledge were also a significant influence on decisions. Researchers indicated that *experts* and *researchers/academics* exerted the most influence on their decisions. Policy-makers answered that *experts* were influential (2nd choice). Third, the users of technology also exerted a significant influence on these decisions. Both researchers and business R&D&I leaders indicated that *users* are a relevant influence on their decisions (ex aequo second choice).

In sum, the results suggest that the most important sources of influence on decisions regarding technological innovation were hierarchies, sources of knowledge, and users of technology, although in different combination for each group. (1) The links to hierarchies were emphasized by policy-makers and by business R&D&I leaders; (2) the links to knowledge were relevant to researchers and, to a lesser extent, policy-makers; (3) the links to users were relevant to researchers and business R&D&I leaders. These three patterns suggest that there are some common sources of influence on these three groups. It might be that these common sources of influence also depend on the type of technology decision. The following section will deal with this issue.

Table 5 Number and percentage of answers in relation to the use of indicators by type of technology decision

	Yes		No		Total	
	Answers	%	Answers	%	Answers	%
Buying equipment/ technology	31	79	8	21	39	100
Development of products/technology	24	83	5	17	29	100
Intellectual property rights	1	50	1	50	2	100
Policy design	20	95	1	5	21	100
Total	76	84	15	16	91	100

4.3 Influence of the Types of Technology Decision

The type of technology decision can influence the way a decision is made. In fact, the need for indicators is expected to vary according to the type of technology decision along the lines previously described. Table 5 presents results on the use of indicators by type of decision.

Table 5 reveals that the use of indicators varies in the different types of technology decisions. In fact, 79 % of those buying equipment or technology reported using indicators, followed by 83 % of those who decided to develop products or technology and 95 % of those who were involved in policy design. The number of responses for intellectual property rights is too low to permit analysis. Nevertheless, the results suggest that the type of technology decision influences the use of indicators, as suggested in Sect. 2.

Furthermore, it was suggested in Sect. 2 that the main drivers of the different needs for using indicators were related to the complexity and uncertainty associated with the decision. There is a close association between the type of technology decision in each group suggesting that these drivers are relevant, although a definitive answer requires further qualitative results beyond of the scope of this paper. Table 6 summarizes the types of decisions identified by group in the questionnaire.

The table reveals that the vast majority of those in the researchers group (87 %) decided about buying equipment/technology. The majority of the business R&D&I personnel (69 %) decided about the development of products/technology. As expected, the vast majority of policy-makers (88 %) decided about policy design. Very few decided about intellectual property rights, and no other types of technology decisions were detected in the questionnaires. Therefore, it can be stated

Table 6 Number and percentage of answers to the question "What type of decision did you make in relation to the adoption and/or investment in technology (choose the most relevant one to your actual or past activity)", by group

	Acquisition of equipment/ technology		Development of products/ technology		Property rights		Policy design		Total	
	Answers	%	Answers	%	Answers	%	Answers	%	Answers	%
Researchers	27	87	3	10	1	3	0	0	31	100
Business R&D&I	10	28	25	69	1	3	0	0	36	100
Policy-makers	2	8	1	4	0	0	21	88	24	100
All groups	39	43	29	32	2	2	21	23	91	100

that groups' answers are concentrated around specific types of technology decisions.

In sum, there are three main reasons that the types of technology have an impact on the degree to which indicators are needed. First, the results suggest that some types of decisions depend on indicators more than others. Policy design requires more indicators than the development of a product or technology and, to a larger extent, than the acquisition of equipment or technology. Second, each group tends to make the same type of relevant technology decision. Consequently, each group tends to have different needs for indicators: Policy-makers tend to need more indicators than business R&D&I leaders and even more than researchers (who tend to be the group that needs fewer indicators). Third, results for the type of technology match the results for the use of indicators presented in the first question, where policy-makers are revealed to use more indicators than business leaders and even more than researchers. We can, therefore, conclude that the type of technology decision is a significant factor affecting the use of indicators in decisions.

4.4 Conclusions

The initial assumption that innovation groups use indicators to make technology decisions was confirmed. However, corresponding expectations about the influence they exert on decisions were not confirmed for three main reasons. First, the majority of policy-makers and business R&D&I leaders considered indicators to be mostly symbolic regarding the decision. Second, social relations were considered more important than indicators in the technology decisions of policy-makers and business groups. These two groups were influenced to different degrees by hierarchies, sources of knowledge, and users. Third, the role of indicators also depended on the different types of technology decisions; however, the indicators proved to exert less influence even in those types of decisions in which indicators were claimed to be used more. Therefore, it can be said that indicators play a less influential role than initially thought.

Two different profiles of decisions were identified suggesting that the influence of indicators appears to be contingent to different degrees on the social world and the type of decision. On the one hand, the researchers group used indicators in a more influential way, was less constrained by social relations, revealed abundant relationships with sources of knowledge and users, and decided about the acquisition of equipment or technology. On the other hand, business and policy-makers were relatively more bound to social relations (in increasing degrees) and admitted different influences: The business R&D&I group reported that social relations exerted a moderate but influential role and that the influence of hierarchies and users on their decisions primarily concerned the development of products or technology; and policy-makers reported significant bounds to social relations, with hierarchy and sources of knowledge influencing decisions about the design of policies.

Quantification of the use and influence of indicators is important for two reasons. First, it provides an initial contextualization of the extent to which indicators can be used in decisions, by characterizing the decision-making pattern of those individuals who are most prone to use indicators, namely knowledge-intensive individuals involved in technological innovation. Second, this research also made it possible to detect three predominant approaches dominating an individual's attention during the decision-making: (1) a rational-analytical approach in the case of the instrumental role of indicators, where indicators were used and were also more important than social relations; (2) a political-behavioral approach in case of the symbolic role of indicators, where indicators were used but social relations were classified as more important than the indicators; and (3) the emotional-intuitive approach in the case where indicators play no role at all, i.e., where respondents denied using indicators and revealed that social relations exerted no or only minor influence.[22] The present results show that, first, the decisions of researchers were more strongly based on the rational-analytical approach than those of the other groups (e.g., through stronger influence of indicators). This suggests that their work depends less on social interactions than the other groups and that it was perhaps influenced by the fact that they engaged mostly in simpler acquisitions of equipment or technology. Second, the business R&D&I group moderately emphasized a political-behavioral approach, not only because of their involvement in the development of products or technology where success depends more on social factors, such as the need to interact with company hierarchies and users. Finally, policy-makers significantly emphasized a political-behavioral approach because their decisions in policy design need strong stakeholder involvement and expertise, both significantly dependent on complex negotiation and consensus activities.

---

[22]The results were adjusted to take account of responses to the question of decision-making styles, which is not presented here.

5 Discussion

Indicators are one tool among others that can help during the decision-making process regarding technological innovation. The fact that indicators only played an instrumental role in the minority of decisions suggests that indicators mostly serve as a complementary instrument in decision-making. When used in a relevant manner, indicators can offer support for decision-making, help to confirm a decision or be used as instruments to justify decisions. But there are other significant influences on decisions to be taken into account, such as social relations, the type of decisions or even personal experiences.

The different decision profiles identified in this paper allow for two different lines of discussion. First, the emphasis on the symbolic role of indicators in the business and policy-maker communities appears to suggest that the deliberations of these individuals are more restrained by specific contexts than those of researchers. One difference is that decisions of the former are taken in situations of significant complexity and uncertainty of the development of the products or technology and of the policy decisions, in contrast to the research projects of scientists, who expected to deliver relatively more predictable outputs like papers and books. In fact, policy decisions concerning specific types of technology tend to face complexity and uncertainty from several sources, such as the volatile future of technology markets, job insecurity, and the complexity of the innovation systems. Therefore, decisions taken under these conditions may help explain a larger dependence on social ties and the weaker influence of rational-analytical methods found in the results.

Second, the researchers' decision-making profile inspires a different line of discussion. In fact, in contrast to the other two groups, the answers from researchers suggest that at least their technological innovation decisions tend to be more rational or analytic and perhaps more efficiency-oriented. This difference might be attributed to their relatively more isolated social context and the type of decisions (e.g., mostly concerning the acquisition of products or technology). However, the relative attachment to social relations revealed by the business and policy-maker groups does not necessarily imply kinship and lineage behavior or decisions based on less efficiency. In fact, the attachment to social relations might only reveal some dependence or the need for consultation and/or consensus in order to be able to make reflexive judgments under uncertain conditions.

Last, it is important to take into consideration that the present quantitative findings can be tied to the situation in Portugal, particularly among policy-makers where different social and contextual settings might influence the behavior of actors dealing with indicators. Furthermore, being able to compare their situation with that of other groups outside innovation might help us to understand the uncertainty under which decisions concerning technological innovation appear to be made in some groups. Consequently, further qualitative research should be conducted to complement the quantitative results presented in this work and to analyze remaining questions.

Acknowledgments The author gratefully acknowledges the financial support of the Portuguese Fundação para a Ciência e a Tecnologia (Ref: SFRH/BD/76200/2011), and the comments of the editors on previous versions of this article.

References

Aram E, Noble D (1999) Educating prospective managers in the complexity of organizational life. Manag Learn 30(3):321–342

Baruch Y (1999) Response rate in academic studies—a comparative analysis. Hum Relat 52 (4):421–438

Boavida N, Moniz AB, Laranja M (2013) Indicators for technology management: how innovation groups decide? 1st international conference on public policy. ICCP, Grenoble, pp 1–27

Böhle F (2011) Management of uncertainty—a blind spot in the promotion of innovations. In: Jeschke S, Isenhardt I, Hees F, Trantow S (eds) Enabling innovation. Springer, Berlin, pp 17–29

Carbonell P, Rodríguez-Escudero AI (2009) Relationships among team's organizational context, innovation speed, and technological uncertainty: an empirical analysis. J Eng Technol Manag 26(1–2):28–45

Chapman R, Hyland P (2004) Complexity and learning behaviors in product innovation. Technovation 24(7):553–561

Cray D, Mallory GR, Butler RJ, Hickson DJ, Wilson DC (1991) Explaining decision processes. J Manag Stud 28(3):227–252

Curseu PL, Schruijer SGL (2012) Decision styles and rationality: an analysis of the predictive validity of the general decision-making style inventory. Educ Psychol Meas 72 (6):1053–1062

Dean JW, Sharfman MP (1993) Procedural rationality in the strategic decision-making process. J Manag Stud 30(4):587–610

Dill WR (1975) Public participation in corporate planning—strategic management in a kibitzer's world. Long Range Planning 8(1):57–63

Dosi G (1982) Technological paradigms and technological trajectories. Res Policy 11 (3):147–162

Eisenhardt KM, Bourgeois LJ (1988) Politics of strategic decision making in high-velocity environments: toward a midrange theory. Acad Manag J 31(4):737–770

Fan W, Yan Z (2010) Factors affecting response rates of the web survey: a systematic review. Comput Hum Behav 26(2):132–139

Fioramonti L (2014) How numbers rule the world—the use and abuse of statistics in global politics. Zed Books, London

Freeman C (1995) The "National System of Innovation" in historical perspective. Camb J Econ 19(1):5–24

Freeman RE, Reed DL (1983) Stockholders and stakeholders: a new perspective in corporate governance. Calif Manag Rev 25(3):88–106

Fusari A, Reati A (2013) Endogenizing technical change: uncertainty, profits, entrepreneurship. A long-term view of sectoral dynamics. Struct Change Econ Dyn 24:76–100

Galotti KM, Ciner E, Altenbaumer HE, Geerts HJ, Rupp A, Woulfe J (2006) Decision-making styles in a real-life decision: choosing a college major. Personality Individ Differ 41(4):629–639

Gray B, Ariss SS (1985) Politics and strategic change across organizational life cycles. Acad Manag Rev 10(4):707–723

Grupp H, Mogee ME (2004) Indicators for national science and technology policy—their development, use, and possible misuse. In: Moed HF et al (eds) Handbook of quantitative science and technology research. Kluwer Academic, Dordrecht, pp 75–94

Grupp H, Schubert T (2010) Review and new evidence on composite innovation indicators for evaluating national performance. Res Policy 39(1):67–78

Gudmundsson H, Sørensen CH (2012) Some use—little influence? On the roles of indicators in European sustainable transport policy. Ecol Indic 35:43–51

Hall PA, Lamont M (2013) Why social relations matter for politics and successful societies. Annu Rev Polit Sci 16(1):49–71

Harren VA (1979) A model of career decision making for college students. J Vocat Behav 14 (2):119–133

Hunt RG, Krzystofiak FJ, Meindl JR, Yousry AM (1989) Cognitive style and decision making. Organ Behav Hum Decis Process 44(3):436–453

Ilori MO, Irefin IA (1997) Technology decision making in organisations. Technovation 17 (3):153–160

Jalonen H, Lehtonen A (2011) Uncertainty in the innovation process. Proceedings of the 6th European conference on innovation and entrepreneurship. Academic Publishing, Aberdeen, pp 486–492

Jauch L, Glueck W (1988) An invitation to strategic management. In: Jauch L, Glueck W (eds) Business policy and strategic management. McGraw-Hill, Singapore, pp 41–42

Kemp R, Schot J, Hoogma R (1998) Regime shifts to sustainability through processes of niche formation: the approach of strategic niche management. Technol Anal Strat Manag 10(2):175–198

Kuhlmann S, Shapira P, Smits R (2010) Introduction. A systemic perspective: the innovation policy dance. In: The theory and practice of innovation policy. An international research handbook. Edward Elgar, Cheltenham, pp 1–22

Linn K, De Man AP, Bossink B (2013) The strategic decision making perspective: how do high-tech firms reach decisions? Paper presented at the 35th DRUID Celebration Conference 2013, Barcelona, Spain

Linstone HA (2008) Soldiers of reason: the RAND corporation and the rise of the American Empire, Alex Abella, 2008, Harcourt Inc. (2008) Technol Forecast Soc Chang 75(9): 1451–1453

MacRae D (1985) Policy indicator: links between social science and public debate. University of North Carolina Press, Chapel Hill

Maddison A (2001) The world economy—a millennial perspective. OECD, vol 1. Development Centre Studies. OECD Publishing, Paris

Maddison A (2003) The world economy—historical statistics. OECD, vol 1. Development Centre Studies. OECD Publishing, Paris

Meijer ISM, Hekkert MP, Koppenjan JFM (2007) The influence of perceived uncertainty on entrepreneurial action in emerging renewable energy technology; biomass gasification projects in the Netherlands. Energy Policy 35(11):5836–5854

Menzel S (2013) Are emotions to blame? The impact of non-analytical information processing on decision-making and implications for fostering sustainability. Ecol Econ 96:71–78

Merry SE (2011) Measuring the world indicators, human rights, and global governance. Curr Anthropol 52(S3):S83–S95

Misuraca G, Codagnone C, Rossel P (2013) From practice to theory and back to practice: reflexivity in measurement and evaluation for evidence-based policy making in the information society. Gov Inf Q 30:S68–S82

Morgan DL (2008) Snowball sampling. In: Given LM (ed) The SAGE encyclopedia of qualitative research methods. SAGE Publications, Thousand Oaks, pp 816–817

Murray E, Pollack L, White M, Lo B (2007) Clinical decision-making: patients' preferences and experiences. Patient Educ Couns 65(2):189–196

Narayanan VK, Fahey L (1982) The micro-politics of strategy formulation. Acad Manag Rev 7(1):25–34

Nelson RR, Winter SG (1977) In search of a useful theory of innovation. Innov Econ Change Technol 6:36–76

Nutt PC, Wilson DC (2010) In: Nutt PC, Wilson DC (eds) Handbook of decision making. Wiley, Chichister

Oxford Dictionaries Online (2014) Complexity. In: Oxford Dictionaries Online. Oxford Dictionaries

Palys T (2008) Purposive sampling. In: Given LM (ed) The SAGE encyclopedia of qualitative research methods. SAGE Publications, Thousand Oaks, pp 698–699

Policastro E (1999) Intuition. In: Encyclopedia of creativity, vol 2. Academic Press, San Diego

Porter TMM (1995) Trust in numbers—the pursuit of objectivity in science and public life. Princeton University Press, Princeton

Power M (1997) The audit society: rituals of verification. Oxford University Press, Oxford

Robichaud L, Durand PJ, Ouellet JP (2006) Quality of life indicators in long term care: opinions of elderly residents and their families. Can J Occup Ther 73(4):245–251

Rogers EM (2003) Diffusion of innovation, 3rd edn. The Free Press, New York

Rycroft RW (2007) Does cooperation absorb complexity? Innovation networks and the speed and spread of complex technological innovation. Technol Forecast Soc Change 74 (5):565–578

Sainio LM, Ritala P, Hurmelinna-Laukkanen P (2012) Constituents of radical innovation—exploring the role of strategic orientations and market uncertainty. Technovation 32 (11):591–599

Sauermann H, Roach M (2013) Increasing web survey response rates in innovation research: an experimental study of static and dynamic contact design features. Res Policy 42 (1):273–286

Saumure K, Given LM (2008) Nonprobability sampling. In: Given LM (ed) The SAGE encyclopedia of qualitative research methods. SAGE Publications, Thousand Oaks, pp 563–564

Schoemaker PJH, Russo JE (1993) A pyramid of decision approaches. Calif Manag Rev 36 (1):9–31

Schutt RK (2008) Investigating the social world: the process and practice of research, 6th edn. Pine Forge Press, Los Angeles

Schwartz B, Ward A, Monterosso J, Lyubomirsky S, White K, Lehman DR (2002) Maximizing versus satisficing: happiness is a matter of choice. J Pers Soc Psychol 83 (5):1178–1197

Scott SG, Bruce RA (1995) Decision-making style: the development and assessment of a new measure. Educ Psychol Measur 55(5):818–831

Sheldon EB, Parke R (1975) Social indicators: social science researchers are developing concepts and measures of changes in society. Science 188(4189):693–698

Shih TH, Fan X (2009) Comparing response rates in e-mail and paper surveys: a meta-analysis. Educ Res Rev 4(1):26–40

Simon HA (1959) Theories of decision-making in economics and behavioral science. Am Econ Rev 49(3):253–283

Simon HA (1979) Rational decision-making in business organizations. Am Econ Rev 69 (4):493–513

Six P (2002) Can policy making be evidence-based? MCC: Building Knowledge for Integrated Care 10(1):3–8

Smith M, Higgs J, Ellis E (2008) Factors influencing clinical decision making. In. Higgs J, Jones MA, Loftus S, Christensen N (eds) Clinical reasoning in the health, pp 89–100

Smits R, Van Merkerk R, Guston DH (2010) The role of technology assessment in systemic innovation policy. In: Smits R, Kuhlmann S, Shapira P (eds) The theory and practice of innovation policy: an international research handbook. Edward Elgar, Cheltenham and Northampton

Starcke K, Brand M (2012). Decision making under stress: a selective review. Neurosci Biobehav Rev 36(4):1228–1248

Sull D, Eisenhardt KM (2012) Simple rules for a complex world. Harv Bus Rev (September):1–8

Swami S (2013) Executive functions and decision making: a managerial review. IIMB Manag Rev 25(4):203–212

Thunholm P (2004) Decision-making style: habit, style or both? Personality and Individual Differences 36(4):931–944

Turpin SM, Marais MA (2004) Decision-making: theory and practice. ORiON 20(2):143–160

Utterback JM (1974) Innovation in industry and the diffusion of technology. Science 183 (4125):620–626

Van de Ven A, Poole MS (1990) Methods for studying innovation development in the Minnesota Innovation Research Program. Organ Sci 1(3):313–334

Waelbroeck P (2003) Innovations, production complexity and the optimality of R&D. Econ Lett 79:277–282

Wonglimpiyarat J (2005) Does complexity affect the speed of innovation? Technovation 25 (8):865–882

Part II
Extra-Disciplinary Transgression of Boundaries

Shifting Practices of Academia as an Entrepreneurial Organization in Indonesia

The Case of ATMI Polytechnic Cikarang

Farah Purwaningrum

Abstract

The study examines how the polytechnic—as a knowledge producing organization in the Indonesian science system—produces and exchanges knowledge with other organizations in Indonesia. The study is situated in organizational sociology. The term "knowledge" in this chapter refers to tacit knowledge produced by knowledge-based workers in the polytechnic field. It discusses the production-based education method as well as entrepreneurial and academic organization. Empirically, it demonstrates the organizational change process heading towards the market and how collaboration between polytechnics and industry is achieved. By using reflexive ethnography, the chapter shows how practice at ATMI Polytechnic Cikarang Indonesia is changing, with the actual orientation of work shifting to an entrepreneurial organization.

Keywords

Production-based education · Polytechnic · Indonesia · Knowledge sharing · Organizational change

This chapter addresses the question of how the internal structure of modern academia is affected by today's knowledge-based economy and the increasing liberalization of the education sector. Governments in Southeast Asia, Indonesia and Malaysia in particular (Evers et al. 2010; Purwaningrum 2014a) are increasingly

F. Purwaningrum (✉)
Brunei Muara District, Brunei
e-mail: farah.purwaningrum@ubd.edu.bn; fara.arum@gmail.com

© Springer Fachmedien Wiesbaden GmbH 2016   103
B.-J. Krings et al. (eds.), *Scientific Knowledge and the Transgression of Boundaries,* Technikzukünfte, Wissenschaft und Gesellschaft / Futures of Technology, Science and Society, DOI 10.1007/978-3-658-14449-4_5

paying more attention to research with links to industrial applications. Preliminary research in the knowledge hub of Brunei Darussalam indicates that despite the minimal linkage of industries with local universities, there is linkage between information communication technology (ICT) industries and polytechnic institutions. The study of industrial clusters in Penang, Malaysia, points out that adequately trained workforces, with appropriate vocational skills, are required to enhance the potential of the industrial sector (IPPTN 2010). However, the role of professionals in the vocational sector is often ignored (Grollmann and Rauner 2007). Polytechnics have been overlooked in most scientific work from academia, including work conducted in Indonesia, yet there is a growing consensus that polytechnics and universities are important institutions within the sphere of higher education (Tjakraatmadja et al. 2008; Yuliar and Syamwil 2008).

How is knowledge produced in the polytechnic field? Furthermore, how is this knowledge exchanged within the technical industry? These are the questions guiding the discussion in this chapter, which will focus on examining the process of organizational change. The study is situated in organizational sociology in looking at how a polytechnic—as a knowledge producing organization in the Indonesian science system—produces and exchanges knowledge with other organizations. I adopt a broad definition of "academia" in this chapter, which includes polytechnic institutions, universities, and R&D institutes.[1] To begin, academia is a part of the Indonesian science system; the organizational study presented in this analysis is focused on the case of ATMI Polytechnic Cikarang, Indonesia. The study is drawn from reflexive ethnography[2] in its data collection and analysis (Burawoy 1998). Thus, the field is the starting point. Not only did I observe the everyday activities of parts of manufacturing or teaching, but I also observed the relationships in the structures patterning interactions, for example in the job

---

[1]Hence, I treat these knowledge producing organizations on an equal level for the purpose of sociological analysis. In addition, I start from the context, which in this case is the Indonesian context that situates these organizations on the same page. Conceptually they are different organizations. Polytechnics, as a part of the vocational education system, put more emphasis on the development of competence as human capital for the purposes of occupational work (see Fischer and Boreham 2008a). Universities, on the other hand, are often tied with the idea of an "ivory tower" and the production of scientific knowledge.

[2]The implications of using an extended method in the overall rubrics of analysis in the organizational research are: *Firstly*, one will encounter a dialogue form of discussion between the researcher and the respondents; *secondly*, critical thinking: I often asked myself whether the rendition or analysis of data is accurate, whether the questions I have asked suffice and are sufficient as a follow up. *Thirdly*, I acknowledge that my entrance into the field is also influenced by who I am. I am a female Indonesian who speaks fluent Indonesian and basic Javanese.

order or subcontracting given by multinational companies. The data collected includes 13 semi-structured interviews, descriptive statistical inferences from 33 questionnaires (from a total of 47 lecturers and instructors) and participant observation. Participant observation was conducted through an internship in the Polytechnic from 16 November 2010 to 17 February 2011.

Before continuing with the discussion, a clarification of the terms used is required. The term "knowledge" in this chapter will refer to the tacit knowledge[3] (Polanyi 2009) produced by knowledge-based workers in the polytechnic field. In this sense, knowledge is a valuable resource in the area of production (and in the manufacturing process) (Bell 1999; Menkhoff et al. 2011). Additionally, the term "entrepreneurial" refers to an entity such as an organization which incorporates the practices and values applicable in the manufacturing sector (as part of the market) within the polytechnic field. My definition of what entrepreneurial is, thus, differs from the definition of entrepreneurship in, for example, the study carried out by Peterson and Berger (1971).[4]

The study shows how practice at ATMI Polytechnic Cikarang is shifting, with the actual orientation of work heading towards that of an entrepreneurial organization. It bears a resemblance to a supplier company. One needs to take an empirical look at how organizations such as polytechnic institutes produce knowledge and how this in turn affects the collaboration between academia and industry. For example, ATMI Polytechnic survives on the basis of being entrepreneurial in its function, as it produces tacit knowledge concerning parts manufacturing. This manufacturing orientation, along with its socially-embedded relationship with the community, enables collaboration between the Polytechnic and industries.

This chapter is divided into the following sections. The first section discusses the production-based education method in Indonesia's vocational education system. The second section investigates the conceptualization of entrepreneurial organization and academic organization. The third section examines the first shift of tacit knowledge

---

[3]For a conceptual and empirical discussion of tacit knowledge, please refer to the third section of this chapter. I describe the first shift of tacit knowledge in manufacturing processes in this section.

[4]Peterson and Berger's study (1971) was based on the record industry and how it coped with turbulent market environments. It captured the three organizational strategies whereby organizations may adapt to entrepreneurship; first by segregating the environment-linked segment, second by isolating the entrepreneurial function, and third by limiting entrepreneurial liability. The study is cited due to its similarity of focusing on how an organization survives by greater integration into a market environment. Yet, in contrast to their study, I present herewith how the shifts or changes of organizational orientation have taken place in an intended or unintended manner.

production for the purpose of production. The fourth section discusses the second shift of integration further towards the market and how this has had an impact on the way a polytechnic works. The collaboration between a polytechnic and industry is the main focus analyzed in the fifth section. The last section sums up the discussion and provides issues for further research in the theme of organizational change.

1 Production-Based Education Method in Indonesia's Vocational Education System

The existing literature discussing vocational education in Indonesia is relatively limited in terms of explaining the process of internal knowledge production in an organization and in terms of examining collaboration between academia and industry in Indonesia. Wilson (1991), for instance, analyzed technical and vocational organizations in Indonesia and Malaysia. Focusing on the *Sekolah Tehnik Menengah* (or vocational high school), he conducted an analysis using time series data based on statistics from 1985–1986 and archival analysis of planning documents (Wilson 1991). He stated that Indonesia is in the lead compared to Malaysia in terms of the degree of interaction and involvement in curriculum reforms between local companies and technical and vocational academic systems. This is possibly due to the complex Dutch heritage in the academic system in Indonesia (Wilson 1991). On another level of reviewing the progress of the vocational academic systems in Asia, Tilak (2002) argued that Indonesia (along with other countries, namely Malaysia, Philippines, Thailand, and Sri Lanka) has "moderately developed" vocational and technical academic systems. Yet these studies focus more on access to academia and curriculum building rather than on knowledge management and linkages with external organizations. Organizational sociology has not been the focus of analysis in vocational education in Indonesia, especially in noting the changes arising due to the openness towards the market system and values offered from a production-based education system.

The application of a production-based education method (often phrased as production-based academic and training system) in Indonesia's vocational education system is still rare. Generally, the higher education policy in Indonesia situates polytechnics on a par with traditional universities. This is apparent in the fact that polytechnics in Indonesia have to carry the *Tri Dharma* functions, which were originally designed for universities. They include teaching (*pengajaran*), research (*penelitian*), and community service (*pengabdian masyarakat*) (Buchori and Malik 2004). Situating polytechnics akin to universities may cause problems in terms of professional certification, as lecturers in polytechnics are obliged to do research as part of their function, similar to university-level lecturers. The policy has a macroscopic view in the sense that it neglects the value of a production-based education system. Indeed, rather than receiving research funding, production-based academic systems support themselves through production activities, such as accepting job orders (subcontracting) from industries. Hence, the majority (almost 80 %) of the costs of education are subsidized by income from production activities (Jacobs-Foundation 2009). According to the latest data in 2013 from the Directorate General of Higher Education—Ministry of Education in Indonesia (DIKTI), there are more than 200 polytechnics in Indonesia, of which some 32 are funded primarily by the state (DIKTI 2012).[5] However, to the author's knowledge, up to now there are only two polytechnics that rely directly on market subsidies to cover operational costs. These two institutes are ATMI Polytechnic Cikarang and ATMI Polytechnic Solo. Organizationally, both polytechnics have similar structures in regards to having production-based education systems. At each, there are two directorates that play a significant role in the functioning of a production-based education system. They are the production directorate and the academic directorate. Yet, these two polytechnics are unique in that production or, to be more exact, manufacturing of parts[6] has become the central focus. As discussed later, job orders are a prominent channel of knowledge production in ATMI Polytechnics, to the extent that concentrated learning takes place in the production directorate and does not flow to the academic directorate. Moreover, distinct from other knowledge-producing organizations in the Indonesian science system (in this case R&D institutes and universities, which are dependent on state funding; see: Purwaningrum 2014b), the majority of operational costs are covered by the subcontracting activities of polytechnic institutes. Hence, apart from carrying out teaching functions, the

---

[5]See the following website for further information: http://forlap.dikti.go.id/ (accessed 18 May 2015). PDPT is the portal for the higher education database (or *Pangkalan Data Pendidikan Tinggi*).

[6]"Part" as a term in this chapter is used to refer to goods produced in the assembling mode, such as moulds, blow moulds, die casting, dies, special purpose machines, and jig fixtures. It also includes component spare parts, such as shafts, gears, flanges, and parts composed of ferro- as well as non-ferro-based materials (email communication with an instructor, 3 July 2011). This part can be a complete part whereby no further process is required (*barang jadi*), or an incomplete part whereby further processing and additional manufacturing is necessary (*barang setengah jadi*) (informal discussion with an instructor, Cikarang, 2 Feb. 2011). A common stand is that the production of these parts is based on the needs and orders made by industries. These practices of manufacturing and producing parts are denoted as job orders (interview with an instructor, Cikarang, 17 June 2010).

polytechnics also resemble supplier companies in a supply chain. In this sense, they position themselves as supplier companies in the manufacturing market. Studying the organizational model of a production-based education system as a model of financing academia may be worth exploring as an alternative funding method amidst a climate of underfunding in the Indonesian science system.

A brief overview of the history of ATMI Polytechnic is warranted prior to going further in the discussion. The history of both polytechnics started with the ATMI Polytechnic Solo, Indonesia, which was opened in 1968 by three Jesuits from Switzerland: *Romo*[7] Gerard Chetelat, *Romo* Paul Ammann, and *Romo* Johann Balthasar Casutt (Teiseran 2010). The idea for ATMI Polytechnic in Solo actually began in 1963 (Casutt 1991) with the help of the Franz Xaver Foundation in Zurich and was made possible by the cooperation of the Swiss Ambassador (Triatmoko 2009; Teiseran 2010). Upon receiving financial assistance from the Swiss Government and the Franz Xaver Foundation, the campus of ATMI Polytechnic in Solo was built in Karangasem Village. ATMI Polytechnic in Cikarang is a formally separate entity from the ATMI Polytechnic in Solo. The ATMI Polytechnic in Cikarang was established in 2001 by the same organization as that in Solo, namely the Karya Bakti Foundation, which is a Catholic organization of Indonesian Jesuits (Triatmoko 2009). It started its operational activity after receiving donations from the German government, Indonesia's business conglomerates, and ATMI Polytechnic alumni from Solo (Teiseran 2010).

ATMI Polytechnic Cikarang is located in the Jababeka Industrial Park in West Java, Indonesia. It is one of the largest industrial clusters in Indonesia. Initially, it was created to support manufacturing training and education for workers in Jakarta and Bekasi—part of the Jakarta metropolitan area (Triatmoko 2009).

Both polytechnics, in fact, were opened with the goal of providing education for manpower in the manufacturing sector and training for the Bekasi residents in general. Despite this initiative of providing training for manpower, most students at ATMI Polytechnic Cikarang, however, were not from the Bekasi area. In an interview with the former director of ATMI Polytechnic in Cikarang, he stated that the majority of the students were from Central Java, Indonesia. In 2010, out of the 27 students from the Jakarta metropolitan area, West Java and Sumatra, very few were from Bekasi, and even they were pendatang (incoming residents) who had not stayed in the area for many years (interview, Cikarang, 17 June 2010). Relations between the ATMI Polytechnic and locals living near the Jababeka Education Park were tense during the beginning of campus construction. One of the Polytechnic's

---

[7]*"Romo"* is a Javanese term; in a Catholic-based education system, it is synonymous with the term "Father".

assistant directors offered the locals free training in manufacturing-related areas, yet this was rejected. The locals then opted to take advantage of the waste from the parts produced there, including the materials that were not used, and this in fact yielded a more direct income for them (informal discussion, Nov. 2010). The offer of training, however, still holds to the present time.

The next section will examine two typologies of organizations: The entrepreneurial organization and the academic organization. The discussion in this section will also provide a basis for the empirical analysis.

2 Conceptualizing Organizations: Entrepreneurial Organization and Academic Organization

This section will illustrate two analytical organization formats: Entrepreneurial organization and academic organization. For the purpose of the analysis, I argue that the inquiry of organizational knowledge production and academia-industry collaboration requires two foci of engagement. The *first focus* is the analysis of the epistemic culture (Knorr-Cetina 1999) and the actual work orientation of the organization itself, namely: *How it produces knowledge and how the existing social structure of the organization governs the orientation of production.* Yet, this type of micro-sociological approach has been criticized as an "internalist history of company" (Weingart 1988). However, it does not and should not stop there. The spatial dimension of knowledge needs to be captured in the analysis as well. To that end, the *second focus is how academia responds to the reconfigurations of the notion of space and scale* as one of the main motors of socio-economic transformation (Perry and Harloe 2007). This translates in the field to the notion of not only geographical space, but also social space (Bourdieu 1985, 1989) and cultural space (Meusburger 2008)[8] which individual actors in industries use in line with their shared frame of reference to foster knowledge-flow processes.

Analysis in the discipline of organizational sociology should be more open in observing the process of knowledge production and knowledge sharing in entrepreneurial organizations. This is not only in terms of everyday knowledge production from a constructivist actor's perspective (see, for example, Hornidge 2007), namely in terms of the role of the self and emotionality (Newton and Smith

---

[8]Meusburger (2008, pp. 66–67) sees culture as a system that is not stable. It comprises signs and interpretations that also incorporate processes and take place in a constant motion. The shared frame of reference can be exercised through the usage of a language, such as Japanese, between first tier supplier and second tier suppliers.

2002) that might affect knowledge sharing, or the organizational narratives, which are the main modes of communicating in organizations (Czarniawska 1998), but also in terms of different traditions, such as ethnomethodology (Turner 1970), which should be taken into consideration when observing knowledge-sharing processes in the context of formal and informal meetings.

Studies on entrepreneurial organizations have been carried out in industrialized countries. Entrepreneurial practices adopted by the University of Waterloo, Canada, have been captured in terms of its intermediary function through its Cooperative Education Program, which connects firms directly with students, tapping into experienced researchers who are attuned to the research and development needs of industry, joint research projects, and project-oriented consulting (Bramwell and Wolfe 2008). By defining an entrepreneurial university as a "university that has developed a comprehensive internal system for the commercialisation and commodification of its knowledge", Jacob et al. (2003) examine the impact of entrepreneurship in Chalmers University of Technology in Sweden. Problems range from accommodating values of commercialization of knowledge, ad hoc trial of the Chalmers infrastructure for innovation and entrepreneurship, to the question of whether research should be treated as a public good or if it should be integrated into the sphere of knowledge production with the rest of the economy (Jacob et al. 2003, pp. 1563–1564). In Southeast Asia, particularly in Indonesia, the notion of "entrepreneurialism" in the science system has been dealt with primarily in the context of universities. Nugroho (2012, p. 17), an Indonesian sociologist who reflected on his experience working at Gadjah Mada University in Yogyakarta, Indonesia, for example, refers to the lack of funding from the state for universities as a reason as to why lecturers and researchers look for extra income outside of their daily work. This propels the entrepreneurial character of how lecturers and researchers carry out their professional work (see also Nugroho 2005; Widmer 2007). A common strand in these studies is that polytechnics have not been a main theme of analysis in observing organizational shifts in the context of entrepreneurial organizations. Studies dealing with knowledge production in polytechnics have primarily been conducted through the lens of pedagogy or vocational education. This is exemplified in the case of Germany (Mueskens et al. 2009) and the UK (Boreham 2002). Hence, the analysis presented in this study aims to contribute to the organizational sociological analysis of polytechnics. In doing so, the following paragraphs will revisit some of the conceptual underpinnings[9] of entrepreneurial organization and academic organization.

---

[9]As stated earlier in this writing, I situate universities on par with polytechnics to grasp the idea of entrepreneurial organization.

From a pedagogical perspective, Clark (1998) investigated five universities: Warwick in England, Twente in the Netherlands, Strathclyde in Scotland, Chalmers in Sweden, and Joensuu in Finland. His method of inquiry was based on a dozen, more or less, interviews, review of documents, and observations, where possible, on the campus activities. He portrays higher education institutions as follows: "Pushed and pulled by enlarging, interacting streams of demand, universities are pressured to change their curricula, alter their faculties, and modernize their increasingly expensive physical plant and equipment—and to do so more rapidly than ever" (Clark 1998, p. xiii). He inferred five characters of the universities' transformation in his study (ibid., pp. 5–8): (i) an expanded developmental periphery in which one can see a growth of units including new non-departmental units, (ii) a strengthened steering core that is quicker and more flexible, (iii) a funding base that is diversified, (iv) a stimulated academic heartland whereby a modified belief system is adopted, and (v) an integrated entrepreneurial culture whereby universities develop a work culture-embracing change. The study has been criticized for a combination of reasons. Methodologically, Deem (2001, p. 16) considers that the theoretical framework offered by the case studies that utilized grounded theory lacks clarity with respect to the process of case-study data analysis. With regard to the organizational perspective, there seems to be a monocultural perspective emerging as the sole authentic view derived from those of senior managers (Finlay 2004). The study, however, provides a basis for observing how a transformation by market forces internally affects universities.

A differing sociological (sociology of communication and, generally, of innovation) perspective is the "triple helix" approach. The triple helix mode of innovation brings forth the conceptualization of an entrepreneurial university model. There is increasing linkage between the users of knowledge, such as industries and/or government bestowing a university/polytechnic institute a role as an economic actor (Etzkowitz 2008; Etzkowitz and Zhou 2008). The triple helix approach focuses on the network overlay of communications and expectations that restructure the institutional order between universities, industries, and government agencies (Etzkowitz and Leydesdorff 2000). Thus, it is a more fully informed (Leydesdorff and Dubois 2004) communication process between university industry and government as well as an evolutionary model (Leydesdorff and Etzkowitz 1998). This theory lends support to a functional approach (for a discussion see Luhmann 1995). There may be limitations to this triple helix approach with respect the observation of internal changes in an organization such as academia. Specifically, there is minimal attention given to history[10] and to the micro level of social interaction that may play a role in knowledge production at the organizational level. The approach is valuable and warrants merit in the sense that the linear process of transfer from origin to application of knowledge is no longer considered as decisive (Leydesdorff and Etzkowitz 1998). This means that such a process is only tenable if it incorporates the possibility of non-linearity, i.e., that the application of knowledge could result in a two-way exchange process. Furthermore, as a research matter it opens up the possibility of a system having a self-organizing capacity beyond even government intervention or involvement. Applying such analysis to the linkage between universities or polytechnics with industries in an industrial cluster would be intriguing, as the starting point would be the belief that a cluster would have its own self-organizing capacity to stabilize the different linkages of knowledge flow. This entrepreneurial university/poly-technic model acts as the motor of the triple helix; it is not governed by the government or industry. Etzkowitz (2008) further explicates the norms that drive the entrepreneurial university as consisting of:

> First, *capitalisation*. Knowledge is created and transmitted for use as well as for disciplinary advance; the capitalisation of knowledge becomes the basis for economic and social development and thus of an enhanced role for the university in society. Second, *interdependence*. The entrepreneurial university interacts closely with industry and government; it is not an ivory tower isolated from society. Third, *in-dependence*. The entrepreneurial university is a relatively independent institution; it is not a dependent creature of another institutional sphere. Fourth, *hybridisation*. The resolution of the tensions between the principles of interdependence and indepen-dence is an impetus to the creation of hybrid organisational formats to realise both objectives simultaneously. Fifth, *reflexivity*. There is a continuing renovation both of the internal structure of the university as its relation to industry and government changes and of that of industry and government as their relationships with the uni-versity are revised (Etzkowitz 2008, p. 41).

There is, of course, criticism of the notion of an entrepreneurial university/poly-technic. Giroux (2001, pp. 1–3) contends that higher education must not be reduced to its entrepreneurial function, although public and private organizations should still be able to gain profit. It is necessary to defend higher education to maintain it as an autonomous sphere and for the progression of a critical and

---

[10]As pointed out by Leydesdorff and Etzkowitz (1998), historical analysis is useful to the extent that one can reconstruct on the basis of existing comprehension how the actors can learn to control the prevailing contingencies. They then point out how evolutionary models do not focus on the historical contexts per se, but on the working of the emerging systems of innovation.

dynamic citizenry (Giroux 2001). Thus, empirical research is required to observe the impact of academia becoming more entrepreneurial; whether it would produce more "elitist" knowledge or whether it would become more like a corporation (see Dhliwayo 2010) following an economic logic. I have so far sketched the concept of entrepreneurial organization to conceptually situate the discussion of this chapter. In the following paragraphs, I will explicate the term academic organization.

Academic organization has often been employed as having several unique characteristics. It carries with it the idea of academic freedom. This can be understood as the scholars having the autonomy to pursue truth despite the fact that the findings they have might upset pieties (Fuller 2005). Furthermore, the student or lay public is protected from any obligation to accept the scholar's findings (Proctor 1991; Fuller 2005, p. 30). Professionals having autonomy and control in the organization are also an essential feature. Baldridge et al. (1978, p. 25), for instance, illustrate how it has unclear and contested goal structures; almost anything can be justified and almost anything can be attacked as illegitimate. It provides service to clients who seek input into the decision-making process and, in addition to this, it must serve its clients with technology that is holistic and non-routine and thus problematic. Henceforth according to them, it is an important case of a professionalized organization where professionals serving the clients call for a significant degree of control of the organizations' decision(-making) processes (Baldridge et al. 1978 as cited by Santos et al. 1998). This leads us to the idea of professional organization.

A professional organization can be distinguished from a semi-professional organization in several dimensions. Firstly, there is knowledge, which underpins the authority of professionals. Indeed, the basis of professional authority is knowledge, and the relationship between administrative and professional authority is largely affected by the kind and amount of knowledge owned by professionals (Etzioni 1969, p. xiii). Secondly, there is a goal, whereby in a professional organization the major goal activities are performed by professionals and are under the authority of professionals (Etzioni 1964, p. 86). Academic organization can be categorized as a type of professional organization yet with a multi-purpose designation. It does not only perform teaching, but also research. Specifically in the case of the polytechnic, it carries the purpose of teaching the students but at the same time it discharges the function of production. Etzioni's (1964, p. 3) analysis of professional organization starts from an understanding that organizations are social units that are deliberately constructed and reconstructed to pursue specific goals. This is an understanding he derived from Talcott Parsons' work of *Structure and Process in Modern Societies* (Etzioni 1964, p. 3). It is not only training and academic qualifications that carry weight in the context of knowledge for professionals, but also the actual work can add competence and skills in professionals' knowledge.

Concepts of entrepreneurial organization and academic organization may serve as a continuum to remind one of the shifts that are possible as a result of opening an academic organization to the market to fund the working and functioning of a knowledge producing organization. Later, I will show how a polytechnic is now actively seeking resources directly from the market and how it competes with firms (second tier suppliers) to accrue income. Yet how does this affect the culture of knowledge production within the organization? A study of the culture of knowledge production on an organizational level was carried out by Knorr-Cetina (1999). She demonstrated, among other things, how molecular biology was becoming a dual-layered organization with ensuing struggles bearing the mark of a territorial conflict among units managed by scientific persons (in this case laboratory leaders and individual researchers) competing for resources and success (Knorr-Cetina 1999). Practices of knowledge production located in everyday life, to the extent they are structured and interwoven in an individual's action, need to be integrated into the analysis of organization as well. The context of knowledge production is also spatial (Perry and Harloe 2007; May and Perry 2011). The local or regional level is the site whereby the dynamics of different knowledge capitalisms, from policy conceptualization to conditions of production, are most obvious (May and Perry 2011, p. 133). The process of knowledge exchange itself is, as I will later show, also mediated by spatial context. Having explained these foci of analyses, the paragraphs that follow offer an empirical analysis of how a polytechnic shifts its organizational orientation. The initial shift is in regard to the production of knowledge which is oriented towards manufacturing purposes. The second shift is in regard to the incorporation of manufacturing practices into organizational work. The next section, hence, discusses the former, i.e., how tacit knowledge is produced and the tension that arises due to its production.

3   The First Shift: Tacit Knowledge Production in Manufacturing Processes

Oftentimes types of knowledge are distinguished on the basis of being tacit or explicit knowledge. The two, however, are correlated. Explicit knowledge requires tacit knowledge for it to materialize into action. In Polanyi's words: "Hence, all knowledge is either tacit or rooted in tacit knowledge. A wholly explicit knowledge is unthinkable" (Polanyi 1969, p. 144 as cited in Rammert 2009, p. 275). Studies of organization, especially intelligent organization, have also been carried out from the perspective of management[11] (Willke 1999, p. 4). Nonaka, from a business perspective, contends that dynamic knowledge creation requires a certain degree of externalization and codification of, presumably, tacit knowledge (see Nonaka 1994). This study starts with the definition of tacit knowing as suggested by Polanyi (2009). It incorporates (i) valid knowledge of a problem, (ii) the scientist's capacity to pursue it, guided by his or her sense of approaching its solution, and (iii) a valid anticipation of the yet indeterminate implications of the discovery made (Polanyi 2009, p. 24). I start with this definition as a starting point of my conceptualization of tacit knowledge. But as I will show later on, I expand it further based on my data to incorporate the subjective and objective elements of knowledge.

Discussion of vocational knowledge, especially tacit knowledge, attempts to surpass the concept of tacit knowing coined by Polanyi (2009). Stevenson (2001), for example, pointed out the deficiencies of tacit knowing in Polanyi's conceptualization. He stated how there are considerable problems linked with a strict and exclusive codification of any knowledge. He added that to say one can perform skills but cannot really confer how, as conceptualized by Polanyi, would be inadequate and lack clarity (Stevenson 2001, pp. 655–656). By drawing on de Jong and Fergusson-Hessler's work (1996), he concluded that tacit knowing seems to occupy a central place in the "situational, conceptual, procedural and strategic knowledge of experts" (Stevenson 2001, p. 657). Their critiques are of merit for this study in looking into situational know-how and in unfolding the notion of the organizational level of knowledge.

Knowledge production in the polytechnic field is geared towards the process of manufacturing. I substantiate this postulate by firstly defining knowledge based on the fieldwork conducted on the shop floor in the Polytechnic. I, then, characterize the type of knowledge produced. Based on this, I intend to show that there has been a shift in producing knowledge more towards the interest of manufacturing processes.

Knowledge in the present context is defined as the "know-how on techniques of processing and techniques of assembling" (informal discussion with an instructor, Cikarang, 1 Feb. 2011). This includes know-how of the technical details used in integrating components of produced parts in relation to a larger part (informal discussion with an instructor, Cikarang, 1 Feb. 2011). The typology of knowledge in this area of assembling is similar to the work process knowledge that is vital in

---

[11]In regard to this point, Willke (1999, pp. 4–5) added that a revival of the theme of knowledge society does not come from sociology or political science, but that it instead comes from management theory.

vocational education (Rauner 2005; Fischer and Boreham 2008a, b). Thus, the idea of work process knowledge is to generate a real work environment, such as those used in production processes, for students and at the same time for the process of production.

My observations in the assembling area (Observation, Cikarang, 1 Feb. 2011) portray the subjective and objective elements[12] of this tacit knowledge and further display how instructors and employees demonstrate caution and attention, especially when assembling different parts. They take time in measuring, often using scales and tracing the tolerance of materials. The picture resembles that of a scientist in a laboratory continuously observing the specimen under a microscope when conducting a set of trials. The instructors and employees do not require a manual. These unrelenting exercises and the internalization/embodiment of knowledge (*membadankan pengetahuan*) are guided by "feeling" (interview with an instructor, Cikarang, 17 June 2010). There is no manual, and one learns from continuous and repeated practice, which eventually establishes the feeling:

> The determination of the zero in every drawing is carried out by drawing (design). Yet these machines need feeling, so we could embody ourselves with machine and the manufactured parts (Informal discussion with an instructor, Cikarang, 14 Feb. 2011).

The repeated emphasis on feeling in the discourses (and, as I observed, in action, despite the use of scales) denotes the intuitive character (Schoen 1983) of knowing. It is not solely rational and technical, but also relies on the internal-intuitive process at the individual level. Usually the intuitive level was modelled by a senior expert, who would observe a machine and then conclude that it was not working properly just by listening to its humming. Knowledge in this definition refers to applied know-how regarding the production of parts and would normally be kept tacit yet practiced by means of individual embodiment (*membadankan*), personal memory, and internalization. Thus, it is both the objective and subjective element of knowledge.

---

[12]I use the term "subjective element" to refer to an individual's tacit knowing, which may include intuition. "Objective element" is utilized to allude to an organization's collective way of (unwritten) knowing. The former can be illustrated in the determination of a part/product that does not meet quality requirements, whilst the latter can be seen from the common shared understanding of how to produce a product/part simply by looking at a drawing from an engineer at the Polytechnic.

There are generally two groups of knowledge workers at ATMI Polytechnic, namely the instructors[13] and the lecturers.[14] The instructors at ATMI Polytechnic are engaged in both the routine of teaching students and the manufacturing of industry-related parts (see Fig. 1). In the following passage, an ATMI instructor explains how the two routines are inseparable:

> So in the Polytechnic, we based ourselves on vocation based academic training, students learn from practice. Then they are conditioned in the industrial world whereas in the beginning they learned competence on basic mechanic, then after being deemed as qualified they learn production. What they learn during the competence learning is applied in the production process. This production process is the job order that the Polytechnic receives (interview with an instructor, Cikarang, 8 Dec. 2010).

There are three segments involved in parts manufacturing in the polytechnic field, the first being polytechnic students (notably second-year students) becoming fully involved in production (own observation, Cikarang, 7 Feb. 2011). Secondly, the instructors at the Polytechnic are involved in the teaching of students and at the same time are involved in the manufacturing of parts. Thirdly, there are the employees in the Production Directorate, in particular at the Independent Business Unit (*Unit Bisnis Mandiri*/UBM), Engineering (Design) and Production Planning Inventory Control (PPIC).

---

[13]The instructors are divided to teach students from the first level to the third level. In the first level, the instructors are in charge of teaching the students the basics of technical manufacturing: Engineering including milling, tooling, grinding, drawing, and heat treatment. At the second level, the instructors assist the students in the full production of parts. The second level consists of two cycles (*putaran*). Each cycle includes grinding (*bubut*), milling, learning at center for tools, welding and heat treatment (informal discussion with an instructor, Cikarang, 8 Feb. 2011). The emphasis again is on production. At the third level, the instructors assist students in the production of parts and in training them in different areas, such as pneumatics and electronics. These activities take place mostly in production. The activities of instructors are oriented towards practice rather than theory. The competence of instructors is assessed from an understanding of how to operate machines and produce parts for machines to the required standard of quality (interview with an instructor, Cikarang, 22 Nov. 2010).

[14]The lecturers work within the theoretical realm of teaching. The curriculum for teaching at the Polytechnic is 30 % theoretical and 70 % practical. The lecturers teach the required 30 % component. The lecturers are involved in class-based teaching of various subjects, ranging from material sciences to *Pancasila* (*Pancasila* designates the five values promoted at the state level of ideology in Indonesia). Some of these lecturers are professionals from the industrial sector, such as professionals in the field of progressive dies.

Fig. 1 Instruction given to a student (*Source* author's fieldwork in 2010)

Based on the questionnaire distributed to knowledge workers, I have analyzed the process of learning through parts production that occurs at polytechnic institutes. More than half of the respondents stated that they are involved in the process of parts manufacturing. Of this share, most of the respondents stated that they did learn from being involved in the manufacturing process.

To what extent does learning from experience in parts manufacturing have an impact on organizational routine? The extent of the changes resulting from such learning will be measured by considering two features: (1) the method of teaching and (2) the writing of new scholarly articles.[15] The results show that the changes affect the teaching methods rather than the writing of new scholarly articles. In regard to teaching methods, a large proportion of respondents affirmed that the manufacturing process brings new ideas and changes to the methods of teaching. However, whilst learning from work in production brings changes in the teaching

---

[15]Only 18 respondents are directly involved in the manufacturing process, i.e., lecturers. The remaining 15 are lecturers who are mainly involved in teaching activities at the Polytechnic.

syllabus and methods, there has not been change in new scholarly articles prepared at the Polytechnic institute. Less than half of the respondents stated that the manufacturing process brings ideas and alterations to the publication of new articles.

This indicates that the process of learning is not codified in written form. It is not made explicit by means of the production of scientific knowledge. The fact that learning from work in production brings alterations to teaching methods is prevalent in the workshop. Most of the instructors provide advice and techniques to the students on how to manufacture the parts through direct face-to-face communication.

There is a propensity for instructors to hoard knowledge in a tacit format. I will describe this in the context of codifying knowledge to a teaching syllabus.[16] The respondents that I interviewed stated their concerns regarding the problematic process of codifying learning in the teaching syllabus. One respondent expresses the extent to which the process can be written down and taught in a traditional context; this respondent also emphasized that competence should be the aim:

> I was bewildered by this requirement of teaching syllabus for the second-level instructors. Although I have spent a long time in the Polytechnic, the minute I heard that there should be teaching syllabus for the second level, how will we make it since we are in fact producing parts? What is a teaching syllabus? It is a process if I am correct. Analogous with subjects, physics for example have various formulas, models. It is a competence that needs to be understood and attained (interview with an instructor, Cikarang, Feb. 2011).

The know-how on parts production is tacitly stored; new techniques are kept in one's head, managed through storytelling, not by being written down. Codify something in writing, with the exception of a drawing, is even at odds with the current practice of knowledge production and sharing in the workshop. The next paragraphs will explain the reasons why there is a limit to what can be expressed in a codified manner.

The absence of a standard part and product is the first reason. The nonexistence of a standard part manufactured on a mass-based production line is often given as a cause for the lack of a teaching syllabus and main guidelines for teaching (GBPP). It is difficult to codify this type of learning in a teaching syllabus and in the main guidelines for teaching because the products and parts being manufactured change accordingly with the orders that arise from a particular industry; no part has been standardized yet in the polytechnic field. However, what can be an object of

---

[16]I focus on instructors in the second level who are both teaching students and fully engaged in production.

assessment are one's operating skills and knowledge of machines, as well as the process of learning (informal discussion with the director, Cikarang, 1 Feb. 2011).

The second reason arises from the lack of organizational incentive, especially from the academic directorate. An incentive for achievement is the reward of monetary compensation after an annual performance appraisal. Indeed, what is practiced is very technical and applied. Namely, drawing, assembling and parts production are the daily techniques practiced by instructors (informal discussion with an instructor, Cikarang, Feb. 2011). For instructors, strategic knowledge is acquired in the assembling area, where all the parts are assembled, measured, and checked. This is the technique they master, *the (situated) know-how*. Assembling is situated in the UBM managed by the production directorate. Instructors associate themselves with the production directorate rather than an academic directorate. There is also reluctance to codify the learning in the form of a standard operating procedure (or SOP) for repeated parts/products (informal discussion with an instructor, Jan. 2011). A SOP can be developed at a later date and used as an academic teaching syllabus. I asked one of the key informants about the academic positions supported by DIKTI (Higher Education Directorate—Ministry of National Education), which may be providing monetary incentives for instructors. A respondent explained his lack of interest in the academic position or support from DIKTI. He stated that he cannot be confined by the regulations of academia. His perspective is that the workshop format (*bengkel*) provides more freedom, more knowledge, and more space for instructors to teach (informal discussion with an instructor, Cikarang, Jan. and Feb. 2011).

In response to this perception, efforts are being organized to tackle this issue. A respondent shared his experience of being trained in the *Pekerti* method. Several instructors and lecturers have attended the training. This *Pekerti* training is organized by DIKTI to provide lecturers (*dosen*) with skills relating to teaching (informal discussion with an instructor, Cikarang, Feb. 2011). The *Pekerti* training does not address the issue of changes in parts manufacturing nor does it address the strategy of codifying such learning. Yet another comparative study was initiated by the academic directorate during my internship in order to learn more about codifying learning and know-how in the teaching syllabus. The instructors and lecturers decided to go to the ATMI Polytechnic in Solo:

> Researcher: "Why do you have to go to ATMI Solo to reflect on what you have here for teaching syllabus and main guidelines for teaching (SAP & GBPP)?"
> Respondent: "Well, because we are not yet experienced like them in Polytechnic Solo" (informal discussion with an instructor, Cikarang, Feb. 2011).

After they returned from Solo, I asked what insight was gained in regard to the use (or lack of use) of a teaching syllabus at ATMI Polytechnic Solo:

> The problem in the Polytechnic in Solo is similar with us here. There is no teaching syllabus and main guidelines for teaching (SAP & GBPP). No teaching syllabus for machines, only standard operating procedure. It is difficult to codify the learning process for students, to move from theory to practice or practice to theory (informal discussion with an instructor, Cikarang, Feb. 2011).

Thus, the remedies do not fulfil their assigned tasks. Neither of the efforts has been able to address the issue of codifying the learning and know-how in a teaching syllabus. Consequently, it is the lack of a standard product and of organizational incentives that serve as the main causes of a lack of the codification of know-how into a teaching syllabus. The instructors hoard their knowledge in a tacit format and identify themselves as a part of the production directorate where they can acquire strategic knowledge. The dominant means of producing tacit knowledge as strategic knowledge are used for manufacturing purposes. This shift exhibits the difficulty or the *tension* when the organization cannot convert its function from an entrepreneurial system to an academic entity.

4 The Second Shift: Integration Towards Manufacturing Practices

There is a second shift that is propelling the entrepreneurial feature of a poly-technic. I deal with two issues in discussing manufacturing practices in a poly-technic as an organization, namely concerning knowledge workers and parts manufacturing. The organizational analysis demonstrates how the specialization of their profession poses an arduous challenge to knowledge workers in a polytech-nic. This, in fact, has curtailed the conversion of organizational work from an entrepreneurial objective to an academic agenda. The job order routines demand instructors to take up the role of supervisors apart from teaching. Two complica-tions arise from this setup. First, they have to master a wide array of skills required to fulfil their job. Secondly, due the first complication, it would be a burdensome task for them to achieve specialization. I discussed this issue of specialization with one key informant who is an instructor.

> The job order usually would have been organized by the PPIC. I also teach in the classroom. I am confused though, am I an instructor or a lecturer? I feel more like an industry supervisor. This has impact on my motivation. In here there is no

specialization, such as drawing, or mould or dies or precision parts (informal discussion with an instructor, Cikarang, Feb. 2011).

The measurement for the assessment of performance of lecturers and instructors are the same with the employees. The human resource division assesses performance based on work quality, work quantity, competence, work behavior, independence in working, and initiative in the office, health and safety and cleanliness as well as neatness. The term "competence" is not defined further using this yardstick (interview with a human resource division officer, Cikarang, 10 Jan. 2011). Upon researching specialization, I found there was no satisfying answer provided. Assessment of achievement is based on work performance and is rewarded with monetary incentives (interview with a human resource division officer, Cikarang, 8 Dec. 2010).

The lecturers face challenges in regard to research and development. The emphasis has been on skills: Students' skills or individual skills. In one instance, a lecturer tried to convince the academic directorate of the importance of research, but received a response doubting the function of research, which was the result and influence of on-going activities in the research and development department (informal discussion with a lecturer, Cikarang, Feb. 2011) of a polytechnic institute. A key informant states: "If we do research, what is the reward for us?" (interview with an instructor, Cikarang, 22 Nov. 2010). Currently, there are no credits provided for lecturers when they produce work such as papers or when they update their teaching modules. This is also the case for instructors who perform class-based teaching. Due to the lack of organizational incentives, there is a widespread reluctance to update teaching modules based on tacitly kept know-how (informal discussion with an instructor, Cikarang, Jan. 2011).

Several respondents voiced their concerns. They stated that it is quite burdensome to acquire academic positions at a polytechnic (informal discussion with a lecturer, Cikarang, Dec. 2010 and Feb. 2011). In fact, these academic positions are, at the moment, non-existent. Moreover, there is an emphasis on practical and vocational systems within a polytechnic. The Polytechnic has responded to this need for academic positions and for codifying the teaching process and know-how in the academic teaching syllabus predominantly due to the external demand of accreditation from DIKTI (interview with an assistant director, Cikarang, 17 Feb. 2011). Academic positions have been created as a priority at the Polytechnic starting in 2012 as part of the accreditation process (interview with a human resource division officer, Cikarang, 8 Dec. 2010). The non-recognition of academic positions within the internal structure of human resource management has a significant impact as well. Lecturers at the Polytechnic cannot apply for scholarship opportunities provided by DIKTI, nor can they move up the organizational ladder to achieve academic specialization in a certain field.

What exists in the academic system in the Polytechnic thus is a hierarchical ladder of upward career mobility. A person is readily moving up from instructor, to section head, coordinator of the level/*tingkat*, and then to division head (informal discussion with an instructor, Cikarang, 13 Dec. 2010). The current job evaluation system and job-grading programs also do not capture the horizontal mobility allowing for specialization and ranking of different levels of competence. This exhibits another feature of the academic organization as a hierarchical one. This work organization, which is not completely Taylorist (see Littler and Craig 1978) in character, affects the career structure in the Polytechnic.

The next issue is the organizational setup of parts manufacturing, which is governed by market needs (and thus is profit-oriented). I refer specifically to the usage of the term "non-technical factors", which has been used since the start of my internship in mid-November 2011. This term came up frequently during lunch conversations with key informants or during informal discussions when I was interning in the instructor's office at the workshop.[17]

The expression "non-technical factors" refers to the manufacturing practice governing the parts being produced. The investigation of what constitutes non-technical factors started by asking for the definition as understood by respondents. During one informal conversation in the instructor's office, the following definition was given:

> Non-technical factors are the decision relating to materials/parts for work (*benda kerja*) and the decision regarding the compensation hours for students (if they manufacture faulty parts) (informal discussion with an instructor, Cikarang, 8 Feb. 2011).

Pedagogy ignasiens is one of the non-technical factors affecting parts production. The parts produced are used as an exercise (*latihan soal*) for students. The knowledge required for students regarding which part to manufacture can depend upon the given part's level of difficulty. For example, if a part is too complex, it can be postponed and other relatively simple parts can be produced (informal

---

[17]Documents such as books on ATMI Polytechnic and papers on the flow of parts manufacturing written by the Polytechnic personnel do not address this non-technical factor. The researcher recorded to the best of her ability these numerous informal discussions on the non-technical factors in the field notes.

discussion with an instructor, Cikarang, 8 Feb. 2011). This is consistent with the *pedagogy ignasiens* inherent in the Catholic-based system used in the Polytechnic:

> Well, I can only describe the basics, humanizing humans, so more to *cura personalis*. Each individual is different, so some sort of individual approach is used. When we face groups, for example, we may generalize like a machine but with this (*pedagogy ignasiens*) we can differ the approach, for those students who are a bit slow to learn, or require particular approach, we can personally approach them (interview with an instructor, Cikarang, 8 Dec. 2010).

It has become a rule of thumb that students will be punished for faulty parts (NG or "Not Good") they produce by being "charged" hours in compensation. The number of hours depends on the price of the materials employed. This personal approach also is evident in a case where the instructor had already set hours of compensation for a student for an NG part. The instructor assessed the part in detail, relying on his tacit knowledge.

A good relationship (*hubungan baik*) is also a non-technical factor in successful manufacturing education. A respondent explains how maintaining good relationships with customers is key in the manufacturing of parts (interview with an employee, Cikarang, 14 Feb. 2011). On further investigation into *hubungan baik*, one respondent explained how this works by citing cooperation with an industry whereby a majority of the workers are alumnae of the Polytechnic. This type of relationship can facilitate the *hubungan baik*. In other words, the parts ordered from this actor sharing the same alma mater can be prioritized (informal discussion with an employee, Cikarang, 21 Nov. 2010).

The setting of priorities is the next non-technical factor. Once a part is labelled as top priority or urgent, then there is an issue of lack of time. In this case, there is usually no briefing for students. This has been evaluated by the unit in charge of production at the Polytechnic:

> The NG level has a high percentage during the period of March-December production in 2010. The main cause is due to human error from students and employees. Why are there erroneous mistakes from students? It is due to lack of supervision, tools are limited, so students improvise and end up with NG parts, also due to a lack of detailed briefing, competence of students for machines is insufficient, this is because the basic is from manual machinery, no adequate base in CNC machines (field notes during monthly meeting, Cikarang, 27 Jan. 2011).

I brought up this issue of lack of briefing to an instructor involved in parts manufacturing. He elaborated the reasons for this occurrence:

> Lack of briefing? Oh, that is because we do not have much time in here, and the parts we have to manufacture are urgent. Orders came from PPIC and UBM. I know the

function and tolerance of materials and machines, the standards are fixed. Look, some students in here can work on grinding, some cannot (informal discussion with an instructor, Cikarang, 1 Feb. 2011).

The parts to be manufactured seem to be of such great significance that they might compromise the learning process. The practices of setting priorities for parts would lessen the chance nature of briefing and may be influenced by *hubungan baik*.[18]

It is inevitable that the Polytechnic as an organization learns best through the manufacturing process. When this is the case, then the *pedagogy ignasiens* seem to be compromised and the entire system of the organization functions more like an industry focusing on the output of parts production. However, in those cases when the pressure to produce parts is not so prevalent, the academic facet returns and *pedagogy ignasiens* approaches can be emphasized for the benefit of students.

The manufacturing practices suggest that the intended dual role of the Polytechnic as an industrial and academic organization has shifted. The actual work that the knowledge workers engage in, the difficulty of specialization, and the non-technical factors of parts production all work to give the organization the orientation of an entrepreneurial organization.

At this juncture, then, one has to ask whether the output of the Polytechnic consists in the graduates or the parts. Ideally it should be both. In one discussion where a network is drawn by participants, the student is described as one of the actors with the least influence, but in the end, they become the object. One of the participants came forward and stated the following:

> We often said that the product of the polytechnic is the students. In reality, they are the ones who carry out the system. The parts are the measurement of their competence (*pembanding*). We evaluate the parts they produce. But for the priority in parts manufacturing, they become the object (informal discussion with an instructor, Feb. 2011).

The organization, hence, survives due to its entrepreneurial layer that keeps the machines and the individuals working on the shop floor. It is in this entrepreneurial facet that the Polytechnic nurtures collaboration with companies. This is explained in the next section.

---

[18]This *hubungan baik* or good relation is facilitated by alumni relations.

5 Socially-Embedded Relations: Collaboration with the ISE[19] Company

Knowledge also has a spatial dimension, and the exchange of such knowledge may be governed by social space (Purwaningrum 2013). The ISE Company (or simply put, ISE) is located in the Tangerang, Banten, Indonesia. The Polytechnic itself is located in Jababeka, Cikarang-Bekasi, Indonesia. These two organizations are located around 74 km apart. The parts manufactured by the Polytechnic for ISE are molds. These molds are utilized by ISE for headlights of motorcycles and accessories for equipment associated with the light-lamp system of motorcycles. From the design provided by ISE, engineering staff at the Polytechnic became informed about the development of Honda products, the company for which ISE is a supplier (interview with an instructor, Cikarang, 21 Nov. 2010). It is through the design and drawing of parts that the engineering department at the Polytechnic becomes acquainted with the new models requiring a complex mold. Thus, the drawing and designs are the knowledge obtained from the job order process (interview with an employee, Cikarang, 21 Nov. 2010).

Apart from the design, tacit know-how is acquired by way of engineering visits, mediated through visualization of parts manufactured in the assembling area of the UBM. In most cases, the engineers from ISE would usually visit the Polytechnic to provide advice, or one of the staff of UBM would visit ISE (interview with an employee, Cikarang, 3 Dec. 2010). The interaction and meeting mostly take place in person, not via ICT. Sometimes the discussions with engineers from ISE are at the assembling area and problems in parts are examined. Frequently, the staff from UBM can ask about the dimensions of the mold and the techniques of production, which are not always written in the drawing (observation, Cikarang, 14 and 15 Feb. 2011). Thus, this engineering visit and direct meeting are the key features of the exchange of knowledge between ISE and the Polytechnic.

ISE has assisted the Polytechnic since 2006 in terms of employee training. It took more or less two years for the staff of the Polytechnic to fully comprehend and then be able to work with ISE. In the early years of the consolidation of the Polytechnic in Cikarang, a course in Japanese was offered to the employees. Despite the fact that most of the companies filling the job order are foreign-owned Japanese companies, lack of interest was still an issue. A contrast is the case of German, where most of the employees, instructors and staff speak German, as

---

[19]I can only refer to the company by the initials ISE, as was agreed with respondents during the fieldwork.

some of them have been sent to Fischer A.G. in Zurich, Switzerland, for a one-year internship. In this way, the deficiency in the mastery of Japanese serves as a barrier to accessing know-how (informal discussion with an assistant director, Cikarang, 22 Nov. 2010).

Quite a number of ATMI Polytechnic Solo graduates work at ISE (informal discussion with an employee, Cikarang, 3 Dec. 2010). The respondent then explained in detail how ISE is open for the staff of the Polytechnic to learn, including in *genba* (walk-in process in the shop floor/production plant) and in accessing the shop floor. This bond, forging trust between the Polytechnic and ISE, permits access to learning on the shop floor. The strength of the relationship is indicative of an embedded social relationship, if it is not about the geographical proximity but more about a social space (Bourdieu 1985) enabling the interaction for the knowledge flow process.

The cooperation between ISE and the Polytechnic illuminates the importance of embedded social relations in the exchange of tacit knowledge. The know-how of mold manufacturing is shared using designs, mediated through direct meetings with individuals in a company and visualization of problems or parts in the assembly process. Nonetheless, the lack of mastery of Japanese serves as a barrier to full knowledge sharing. The know-how from ISE is retained in its tacit form, being limited to those who are directly involved in the production process at the Polytechnic, in this case, the engineering division, UBM and instructors at the second level who were involved in the production. Alumni relationships, as demonstrated in this case, show an embedded social relationship that enables the social space of interaction between academia (the Polytechnic) and industry.

6　Concluding Remarks: ATMI Polytechnic Subsumed by Production Orientation

This study is about the organizational change of academia. To sum up, ATMI Polytechnic is an organization that is defined by its orientation towards production and shaped by shifting practices from academic teaching to parts manufacturing. The liberalization of higher education in Indonesia exerts pressure on the survival of academia. The history of ATMI Polytechnic in Cikarang was marked by a Jesuit mission aiming to enable access to education at the vocational level.[20] I argue that one needs to look empirically at the paradigm and structure (of knowledge

---

[20]This is also evident in the process of constructing the new Loyola building in the complex of ATMI Polytechnic in Cikarang. Part of the ceremony is the three *Romo*s (or *Fathers*) and

production) as well as the space enabling collaboration between academia and industry. ATMI Polytechnic attempts to survive despite the overall liberalization of the education sector in Indonesia. It survives by subsidizing its education function by taking on job orders (subcontracting activities).

The organizational shifts towards being more entrepreneurial are framed analytically in terms of the shift in practice concerning the production of tacit knowledge and concerning manufacturing practices internalized in organizational work. In this case, knowledge production is illustrated in everyday interactions, and tacit knowledge is produced on the basis of situated know-how about parts production. Instructors repeatedly promote the internalization of "feeling" or intuition in the mastery of certain skills. This tacit knowledge embodies the subjective and objective elements in this type of field. It is palpable that manufacturing of parts is the main vehicle for the production of knowledge. Tacit knowledge, however, tends to be hoarded. It is not made explicit in the format of a readily-accessible teaching syllabus.

The next shift is in the actual work of the organization, which is geared towards manufacturing. This is identified in two ways, namely the specialization of a profession and the logic of producing parts. In the former, most of the instructors identify themselves as part of the production directorate. This is due to the absence of non-vertical administrative positions, lack of an overall soft-infrastructure to codify tacit knowledge in the teaching syllabus (such as organizational incentives) and standard products. In the latter, the non-technical factors governing parts manufacturing are sketched. These range from *pedagogy ignasiens* and good relations to setting priorities. The second shift sets the manufacturing of parts as a priority, which may compromise the learning process for students.

The interaction with a company, as exemplified by this case study, is fostered by job order requests and pushed by alumni relationships. The embedded social relationship harnessed through an alma mater weave the social space of interaction between academia and industry.

The shift in practices blocks the possibility of such an organization converting from an entrepreneurial organization to an academic one. The more a polytechnic relies on parts manufacturing to survive, as shaped by the everyday practices of the instructors and employees, the more it will be subsumed in being entrepreneurial, forgetting its initial aim of the teaching and academic training of students.

(Footnote 20 continued)
the representative from the funding company holding the four pillars of the future construction on 31 July 2011. This in a way symbolizes partnership.

The organizational analysis presented in this chapter unpacks what emerges at the micro-level in a country that is opening up to liberalization of higher education in a knowledge-based economy. This raises the following issues for future research: What kind of contribution would vocational education bring towards promoting product innovation, especially in light of the difficulty of specialization in a profession in Indonesia? Next, to what extent can the market (re-)orient the type of knowledge being produced by an organization (i.e., academia) meant to produce knowledge to strengthen the knowledge base of an industrial cluster?

Acknowledgments I am sincerely grateful for the reviewers' feedback and the discussions with Solvay Gerke and Hans-Dieter Evers in Bonn, Germany, which have shaped the revision of the chapter. Anastasiya Shtaltovna in Montreal, Canada, has provided detailed and useful feedback for the revision and the overall structure of this chapter.

References

Baldridge JV, Curtis DV, Ecker G, Riley GL (1978) Policy making and effective leadership: a national study of academic management. Jossey-Bass, San Francisco

Bell D (1999) The coming of post-industrial society: a venture in social forecasting. Basic Books, New York

Boreham N (2002) Work process knowledge, curriculum control and the work based route to vocational qualifications. Br J Educ Stud 50(2):225–237

Bourdieu P (1985) The social space and the genesis of groups. Theory Soc 14(6):723–744

Bourdieu P (1989) Social space and symbolic power. Soc Theory 7(1):14–25

Bramwell A, Wolfe DA (2008) Universities and regional economic development: the entrepreneurial University of Waterloo. Res Policy 37:1175–1187

Buchori M, Malik A (2004) The evolution of higher education in Indonesia. In: Altbach PG, Umakoshi T (eds) Asian universities: historical perspectives and contemporary challenges. John Hopkins University Press, Baltimore, pp 249–278

Burawoy M (1998) The extended case method. Sociol Theory 16(1):4–33

Casutt PJ (1991) Die akademi technik mesin industri (ATMI) St Mikail Surakarta. Beruf-bildungssysteme Produktionschule Systemdidaktik. Egebnisse der Fachtagung Gewerbliche Berusbildungprojekte Asien, GTZ, Bandung, pp 1–6

Clark BR (1998) Creating entrepreneurial universities: organizational pathways of transformation. IAU Press and Pergamon, Oxford

Czarniawska B (1998) A narrative approach to organization studies. Sage, Thousand Oaks

de Jong T, Ferguson-Hessler GM (1996) Types and qualities of knowledge. Educ Psychol 31 (2):105–113

Deem R (2001) Globalisation, new managerialism, academic capitalism and entrepreneurialism in universities: is the local dimension still important? Comp Educ 37(1):7–20

Dhliwayo S (2010) The entrepreneurial organisation. In: Urban B (ed) Frontiers in entrepreneurship. Springer, Heidelberg/Dordrecht/London/New York, pp 139–156

DIKTI (2012) The number of polytechnic will be added. http://www.kemdiknas.go.id/
    kemdikbud/berita/293. Accessed 4 Feb 2014

Etzioni A (1964) Modern organizations. Prentice-Hall, New Jersey

Etzioni A (1969) The semi-professions and their organization: teachers, nurses, social
    workers. Free Press, New York

Etzkowitz H (2008) The triple helix university-industry-government innovation in action.
    Routledge, New York

Etzkowitz H, Leydesdorff L (2000) The dynamics of innovation: from national systems and
    "mode 2" to a triple helix of university-industry-government relations. Res Policy
    29:109–123

Etzkowitz H, Zhou C (2008) Building the enterpreneurial university: a global perspective.
    Sci Public Policy 35(9):635–637

Evers HD, Nordin R, Nienkemper P (2010) Knowledge cluster formation in peninsular
    Malaysia: the emergence of an epistemic landscape. ZEF Working Paper Ser 62:1–32

Finlay I (2004) Living in an "entrepreneurial" university. Res Post-Compulsory Educ 9
    (3):417–434

Fischer M, Boreham N (2008a) Occupational work and competence development. In:
    Rauner F, Maclean R (eds) Handbook of technical and vocational education and training
    research. Springer, Dordrecht, pp 439–443

Fischer M, Boreham N (2008b) Work process knowledge. In: Rauner F, Maclean R
    (eds) Handbook of technical and vocational education and training research. Springer,
    Dordrecht, pp 466–474

Fuller S (2005) What makes universities unique? Upating the ideal for an entrepreneurial
    age. High Educ Manag Policy 17(3):17–42

Giroux HA (2001) Introduction. Critical education or training: beyond the commodification
    of higher education. In: Giroux HA, Myrsiaders K (eds) Beyond the corporate university,
    culture and pedagogy in the new millennium. Rowman & Littlefield, Oxford, pp 1–14

Grollmann P, Rauner F (2007) TVET teachers: an endangered species or professional
    innovation agents. In: Grollmann P, Rauner F (eds) International perspectives on teachers
    and lecturers in technical and vocational education. Technical and Vocational Education
    and Training: Issues, Concerns and Prospects, vol 7. Springer, Heidelberg, pp 1–26

Hornidge AK (2007) Knowledge society: vision and social construction of reality in
    Germany and Singapore. LIT Verlag, Muenster

IPPTN (2010) The State of Penang, Malaysia: self-evaluation report. In: Sirat M, Tan C,
    Subramainiam T (eds) OECD reviews of higher education in regional and city
    development. The National Higher Education Research Institute (IPPTN), Penang

Jacob M, Lundqvist M, Hellsmark H (2003) Enterpreneurial transformations in the Swedish
    university system: the case of Chalmers University of Technology. Res Policy 32:1555–
    1568

Jacobs-Foundation (2009) Award ceremony. http://www.4shared.com/mp3/1m08oqPb/in_
    my_arms_-_plumb.htm. Accessed 3 Feb 2012

Knorr-Cetina K (1999) Epistemic culture: how the sciences make knowledge. Harvard
    University Press, Cambridge

Leydesdorff L, Dubois DM (2004) Anticipation in social systems: the incursion and
    communication of meaning. Int J Comput Anticip Syst 15:203–216

Leydesdorff L, Etzkowitz H (1998) The triple helix as a model for innovation studies. Sci Public Policy 25(3):195–203

Littler CR, Craig R (1978) Understanding Taylorism. Br J Soc 29(2):185–202

Luhmann N (1995) Social systems. Stanford University Press, Stanford

May T, Perry B (2011) Social research & reflexivity: content, consequence and context. Sage, Los Angeles

Menkhoff T, Evers HD, Wah CY, Fong PE (2011) Introduction: strategic aspects of developing Asia's knowledge-based economies. In: Menkhoff T, Evers H-D, Wah CY, Fong PE (eds) Beyond the knowledge trap: developing Asia's knowledge-based economies. World Scientific Publishing, Singapore, pp 1–24

Meusburger P (2008) The nexus between knowledge and space. In: Meusburger P, Welker M, Wunder E (eds) Clashes of knowledge. Orthodoxies and heterodoxies in science and religion. Springer, Dordrecht, pp 35–90

Mueskens W, Tutschner R, Wolfgang W (2009) Improving permeability through equivalence checks: an example from mechanical engineering in Germany. In: Tutschner R, Wittig W, Rami J (eds) Accreditation of vocational learning outcomes: European approaches to enhance permeability between vocational and higher education. Nationale Agentur Bildung fur Europa beim Bundesinstitut für Berufsbildung, Bonn, pp 10–33

Newton T, Smith D (2002) Introduction: Norbert Elias and the civilised organization. In: Iterson AV, Mastenbroek W, Newton T, Smith D (eds) The civilized organization: Norbert Elias and the future of organization studies. John Benjamins Publishing, Amsterdam, p 10

Nonaka I (1994) A dynamic theory of organizational knowledge creation. Organization Science 5(1):14–37

Nugroho H (2005) The political economy of higher education: the university as an arena for the struggle for power. In: Hadiz VR, Dhakidae D (eds) Social science and power in Indonesia. Institute of Southeast Asian Studies, Singapore, pp 143–165

Nugroho H (2012) State, university and banality of intellectual: a critical reflection from the inside. Professorial inauguration speech delivered at Gadjah Mada University, Yogyakarta Gadjah Mada University, pp 1–25

Perry B, Harloe M (2007) External engagements and internal transformations: universities, localities, and regional development. In: Scott A, Laske S, Burtscher C (eds) Bright satanic mills, universities, regional development and the knowledge economy. Ashgate, Hampshire

Peterson RA, Berger DG (1971) Entrepreneurship in organizations: evidence from the popular music industry. Adm Sci Q 16(1):97–106

Polanyi M (1969) Knowing and being: essays by Michael Polanyi. The University of Chicago Press, Chicago

Polanyi M (2009) The tacit dimension. The University of Chicago Press, Chicago

Proctor R (1991) Value-free science? Purity and power in modern knowledge. Harvard University Press, Cambridge

Purwaningrum F (2013) The social tie that binds: academia-industry collaboration in ATMI Polytechnic Cikarang, Indonesia. Int J Inf Educ Technol 3(5):547–553

Purwaningrum F (2014a) Knowledge governance in an industrial cluster: the collaboration between academia-industry-government in Indonesia. LIT Verlag, Berlin

Purwaningrum F (2014b) Heading toward knowledge society? Predicaments and challenges of academia in the Indonesian science system, XVIII ISA World Congress of Sociology, Yokohama International Sociological Association

Rammert W (2009) Two styles of knowing and knowledge regimes: between "explicitation" and "exploration" under conditions of "functional specialization" or "fragmental distribution." In: Hage J, Meeus MTH (eds) Innovation, science and institutional change. A research handbook. Oxford University Press, Oxford, pp 256–284

Rauner F (2005) Work process knowledge and development of vocational competence. Oman Conference, Muscat, Oman

Santos F, Heitor MV, Caraca J (1998) Organisational challenges for the university. High Educ Manag 10(3):87–108

Schoen DA (1983) The reflective practitioner: how professionals think in action. Basic Books, Cambridge

Stevenson J (2001) Vocational knowledge and its specification. J Vocat Educ Training 53 (4):647–662

Teiseran M (2010) ATMI from time to time. In: Widiastono TD (ed) Johann Balthasar Casutt SJ: Ab Initio Ad Esse. Bekasi, Tritoenggal, pp 123–133

Tilak JB (2002) Vocational academic training in Asia. In: Keeves JP, Watanabe R (eds) The handbook on educational research in the Asia Pacific Region. Kluwer Academic Publishers, Dordrecht, pp 1–18

Tjakraatmadja JH, Martini L, Wicaksono A (2008) Knowledge sharing in Indonesian context: ITB as potential knowledge hub to create value from academia, business and government linkages. In: 4th International Research Conference on Asian Business: "Knowledge architecture for development: challenges ahead for Asian business and governance". Singapore Management University, pp 1–24

Triatmoko B (2009) The ATMI story: rainbow of excellence. ATMI Press, Solo

Turner R (1970) Words, utterance, and activities. In: Coulter J (ed) Ethnomethodological sociology. Edward Elgar Publishing, Hants, pp 189–207

Weingart P (1988) Close encounters of the third kind: science and the context of relevance. Poetics Today 9(1):43–60

Widmer K (2007) The impact of international projects on Indonesian education and research —a case study of Gadjah Mada University, Indonesia. Freiburger Ethnologische Arbeitspapiere Working Paper 3:1–32

Wilson DN (1991) Reform of technical-vocational academic in Indonesia and Malaysia. Comp Acad 27(2):207–221

Willke H (1999) From artificial intelligence to systemic knowledge management. KI-99: advances in artificial intelligence. Lect Notes in Comput Sci 1701:1–13

Yuliar S, Syamwil IB (2008) Changing contexts of higher education policy: toward a new role of universities in Indonesia's innovation system. IV Globelics Conference, Mexico City, pp 1–10

Designing "Integration Machines"

Computer Simulation and Modeling in Transdisciplinary Sustainability Research in Austria

Judith Igelsböck

Abstract

The paper narrates an Austrian research program in the area of transdisciplinary sustainability research, which strongly inscribes into the promise that bringing together the knowledge and expertise of various (scientific and extra-scientific) actors provides a chance to get a handle on complex societal problems—such as climate change. Starting from the observation that the majority of funded projects makes use of computer modeling and simulation to bring together the knowledge of scientific and extra-scientific actors, the paper aims to understand computer simulation and modeling as "integration machines." Inspired by the way they are presented in the projects themselves in a first place, the notion of the integration machine points to the dynamics of attempts to involve a variety of scientific and extra-scientific actors and the epistemic practices held appropriate to do so. Based on the analysis of the ways how computer simulations and models are discursively designed in different arenas of discussion, development and dissemination (e.g., proposals, publications, interviews, focus groups, project meetings), the paper carves out how "integration machines" incorporate imaginations, hopes and promises, and how they translate between a multiplicity of ascribed attributions. Crucially, the paper attends to different "performative" dimensions of integration machines, showing how they include but also exclude certain kinds of knowledge, how they assume a distinct distribution of responsibilities, and how they (re)produce orders, roles, and identities within the relation between science and society.

J. Igelsböck (✉)
Munich, Germany
e-mail: judith.igelsboeck@tum.de

© Springer Fachmedien Wiesbaden GmbH 2016   133
B.-J. Krings et al. (eds.), *Scientific Knowledge and the Transgression of Boundaries*, Technikzukünfte, Wissenschaft und Gesellschaft / Futures of Technology, Science and Society, DOI 10.1007/978-3-658-14449-4_6

Keywords

Transdisciplinarity · Sustainability research · Participation · Computer simulation and modeling · Integration machines

1 Introduction

Difficulties to adequately respond to pressing and complex societal problems, such as climate change, have posed a challenge to the relations between science and society in the past decades. Criticism has been levelled at the inward orientation of scientific (sub-)disciplines and their disengagement from societal concerns, provoking use of the metaphor of an ivory tower (in which science is trapped) (see, e.g., Shapin 2012). This has also encouraged debate about alternative means of producing knowledge (see, e.g., Hessels and van Lente 2008). These debates are driven by the hope that the multifaceted dynamics of the social, the economic and the ecological spheres can be more thoroughly understood when a wide range of knowledge and expertise is brought together (see, e.g., Borup et al. 2006). More recently, the need for collaboration between scientific and extra-scientific actors (e.g., stakeholders) has in particular been emphasized. By assembling the actors who have a stake in a certain issue, research should become more responsive to societal problems and adaptive to the high degree of uncertainty and complexity that characterizes them (see, e.g., Nowotny 2003).

In this paper, I examine the ways in which calls for collaboration between heterogeneous scientific and extra-scientific actors are translated into integrative research practice. Specifically, I investigate the Austrian research program pro-VISION[1] in the field of transdisciplinary sustainability research, which can be regarded as a prototype of alternative forms of knowledge production as it strongly encourages the participation of heterogeneous—even extra-scientific—actors in the processes of knowledge production. While the research program remains vague on the concrete steps of integration, the majority of funded projects rely on computer modeling and simulation to bring the knowledge and expertise of diverse scientific and extra-scientific actors together. Starting from this observation, I pay attention to the ways in which models and simulations are discursively designed as the central means for producing collaborative knowledge in transdisciplinary sustainability research. By "discursive design" I refer to the arguments and reasons

---

[1]See section "Materials and Methods."

offered for the deployment of computer models and simulations, the ways in which their use is legitimized, the ways in which they are presented as a means for integrating different areas of discussion, development and dissemination (e.g., proposals, publications, interviews, focus groups, project meetings). Following recent work in science and technology studies inspired by participatory design (e.g., Le Dantec and DiSalvo 2013), I relate the notion of "design" to the configuration of roles of diverse participants in transdisciplinary research, as well as to the negotiations about what computer simulations and models are and what they do, or in other words how they can and should be used.

The empirical insights will serve as a basis for the development of the notion of "integration machines". This notion points to the dynamics in attempts to involve a variety of scientific and extra-scientific actors in knowledge production and the epistemic practices employed to achieve such involvement. I will carve out how integration machines are incorporating hopes and promises, and how they are translating between a multiplicity of ascribed attributions. Also, I will point to different "performative" dimensions of integration machines and show how they are including, but also excluding, certain kinds of knowledge. Furthermore, I will discuss how they assume a distinct distribution of responsibilities and (re-)produce orders, roles, and identities within the relationship between science and society.

The aim of this paper is to contribute to a better understanding of the implications of the alternative modes of knowledge production employed at the micro-level. Moreover, it provides insights about the role of computer modeling and simulation in research situations that are characterized by the collaboration of a variety of scientific and extra-scientific actors. It is structured as follows: The nexus between narrations about a new dimension of societal challenges and debates about alternative ways of knowledge production will be discussed first. This will be continued by an introduction of concepts and definitions of transdisciplinarity research. Subsequently, the use of computer models and simulations is related to questions regarding integrative knowledge production. This is followed by an overview of the concrete cases analyzed (namely a range of projects as funded by the above mentioned Austrian sustainability program), and the presentation of my empirical findings. Based on these empirical insights, I will develop the notion of an integration machine. The concluding remarks are devoted to the qualities of this notion for our understanding of the deployment of computer simulations in alternative forms of knowledge production.

2 A New Dimension of Societal Challenges and Alternative Means of Knowledge Production

A new dimension of societal problems has arisen in recent years. This is, in particular, true for the grand challenge posed by sustainable development (see, e.g., Lund Declaration 2009), which is characterized by profound changes in the perception of the relationship between humankind and nature. The notion of the Anthropocene aims to capture the growing awareness "[...] that humankind has become a global geological force in its own right" (Steffen et al. 2011, p. 843). Bruno Latour invokes the term of a "collective *giant*" to grasp the ability of the currently living human population to "[...] *shape the Earth* literally." (Latour 2011, p. 3). With the "[...] growing awareness of human impact on the environment [...]" (Steffen et al. 2011, p. 856), the ways in which we (humans) deal with nature are being reconceived. In the face of pressing problems such as climate change and rather pessimistic prospects for future generations, we can witness a shift from attempts to keep nature under control to the assumption of responsibility for human-environment relationships, which should be translated into collective efforts to manage and govern the relationship between humans and their environment (see, e.g., Steffen et al. 2011).

These expectations of new responsibilities bring complexities that raise questions concerning the potential of disciplinarily organized science and scientific principles, such as the separation between facts and values. In the light of complex situations in which "[...] facts are uncertain, values in dispute, stakes high, and decisions urgent" (Funtowicz and Ravetz 1993, p. 744), scientific autonomy becomes disputable when it comes to identifying problems or assessing the quality of the produced knowledge. As claimed by Latour, "[t]here is no single institution able to cover, oversee, dominate, manage, handle, or simply trace ecological issues of large shape and scope" (Latour 2011, p. 1). This statement can serve as a proxy for the wide agreement to accept contemporary challenges, which will require the collaboration of a variety of actors who have a stake in an issue and are committed to its solution.

A range of concepts of alternative forms of knowledge production has emerged closely entangled to the newly accepted responsibilities towards the environment, such as "mode 2 knowledge production" (e.g., Gibbons et al. 1994) or "post normal science" (Funtowicz and Ravetz 1993). While they certainly differ from each other — for instance, either rather descriptive or prescriptive accounts can be identified (see, e.g., Hessels and van Lente 2008)—they share some key characteristics.

Conceptualizations of alternative ways of knowledge production propose what is called an "extended peer review" (Funtowicz 2001). It can be described as an attempt to integrate actors from outside the scientific community into the production and validation processes of knowledge. This kind of involvement should allow for a new quality of knowledge, called the "social robustness" of knowledge (e.g., Nowotny 2003). Social robustness is connected to an appreciation and inclusion of diverse forms of expertise. Against the background of the impression that disciplinarily produced knowledge often seems disengaged from societal needs, an increased sensibility to the societal contexts of the application of knowledge is being demanded:

> [...R]obustness is tested not only inside the laboratory. The test typically occurs outside the laboratory, in a world in which social, economic, cultural, and political factors shape the products and processes resulting from scientific and political innovation (Nowotny 2003, p. 155).

Yet critique has been voiced—amongst others by Weingart (2008)—that it has remained unclear what exactly these requirements mean and how they can be achieved in practice. In this paper, I want to respond to that critique by scrutinizing the concrete integrative research practices in prototypes of new modes of knowledge production. The focus of this paper lies on a form of knowledge production that is strongly inspired by concepts of alternative ways of knowledge production and has been developed in close relation to ecological issues, namely *transdisciplinarity*. In contrast to interdisciplinarity, transdisciplinary knowledge production is characterized by a "double transgression" of boundaries. First, similar to common definitions of interdisciplinarity, actors stemming from different scientific disciplines are supposed to produce knowledge together. Second—and that is one of the main differences to interdisciplinarity—institutions and people from outside the science systems (e.g., people from non-governmental and governmental organizations, practitioners, regional actors, pupils, and other stakeholders) are also actively involved in the knowledge production processes. In this connection, "actively" means that they are participating in research activities from an early stage on, ideally already in the phase of problem definition (see, e.g., Pohl and Hirsch Hadorn 2008). Transdisciplinarity is thus a form of research in which the boundaries between different scientific disciplines as well as that between science and society are to be transgressed. The integration of extra-scientific actors is specifically aimed to respond to societal demands. Transdisciplinarity, therefore, is defined as

> [...] an extension of interdisciplinary forms of the problem-specific integration of knowledge and methods; while integration refers to scientific questions at the

interface of different disciplines in interdisciplinarity, in transdisciplinarity, on the other hand, it is about integration at the interface of these scientific questions and societal problems (Jahn et al. 2012, p. 2).

As Jahn and colleagues highlight, there is not a consistent definition of transdisciplinarity and also the demarcation to other forms of research is not always that clear.

> Generally the differentiation occurs on the level of cooperation. [...T]ransdisciplinarity thus differs from interdisciplinarity in that it involves cooperation between researchers and 'practitioners' [...] (Jahn et al. 2012, p. 2).

The collaboration of scientists and extra-scientific actors is expected to yield knowledge that can be directly used in societal contexts. Moreover, transdisciplinary knowledge creation should be more open to a diversity of interests at stake than a disciplinary one, and thus can be related to a "democratization of expertise" (see, e.g., Nowotny 2003).

In related work, scholars have dealt extensively with the advancement or evaluation of specific methods of transdisciplinary integration (e.g., Defila and Di Giulio 1999; Bergmann 2003; Max-Neef 2005; Daschkeit 2007). Furthermore, transdisciplinary knowledge creation was addressed from a macro perspective in so far as potential benefits of transdisciplinary research were related to concepts of knowledge or risk societies (e.g., Bunders et al. 2010). I aim to enhance the state of the art of transdisciplinarity by focusing on micro-processes of knowledge production, i.e., the actual epistemic practice (see, e.g., Felt 2010). As introduced above, I am concretely interested in the ways actors stemming from different backgrounds work together, how their cooperation is realized in terms of the apparatuses and methods used, responsibilities distributed, and roles allocated.

3 Computer Simulations and Modeling in Transdisciplinary Sustainability Research

How can knowledge stemming from diverse scientific and extra-scientific actors be integrated? In the transdisciplinary sustainability research projects that I investigated, I found mainly one answer to this question: by means of computer simulation and modeling. Nine of the eleven investigated projects are developing and using computer models and simulations.[2] Considered adequate for an integrative production of knowledge and seemingly meeting the demands and interests of all the (scientific and extra-scientific) research participants involved, computer

---

[2]See also the section "Materials and Methods."

simulations and models dominate the transdisciplinary research designs in the observed projects. Why are computer simulations and modeling so prominent in transdisciplinary sustainability research? What makes them eligible for the collaborative production of knowledge amongst heterogeneous scientific and extra-scientific actors?

In general, models and simulations are not new within scientific procedures. As a consequence of the new possibilities provided by ever more powerful computers, their importance and use has increased considerably, however. That holds at least in certain areas of research, such as climate science, in which "[...] simulation models have become principal means of data collection, prediction and decision making" (Edwards 2010, p. xix), as they make it possible to gain knowledge about global processes that cannot be explored within the confined boundaries of laboratories. While it seems rather easy to differentiate models from simulations,[3] it becomes more difficult to define them in terms of established categories. Peter Galison underlines that "[c]omputers and simulations [...] come to stand in a novel epistemic position within the gathering of knowledge—not quite a piece of empirical machinery, and not quite one with theoretical apparatus" (Galison 1997, p. xix). Related to that normative perspective, other authors have highlighted the hybrid nature of models and simulations. Merz and Hinterwaldner (2012) ascribe them two sides: a representative, or demonstrative, side—such as being of heuristic value, i.e., as a means to understand and learn about the world—as well as a productive one—and as such being a means for intervention. Similarly, Keller (2000) describes how computational models in molecular biology are transgressing the boundary between basic and applied research by simultaneously being "models of"—representing a certain biological phenomenon—and "models for"—being developed to be used. The hybridity of computer models and simulations will also play an important role in transdisciplinary research. It does so in regard to multiple demands placed on models and simulations by diverse scientific and extra-scientific actors, and it will play a role in regard to "boundary work" (Gieryn 1983) as carried out by different scientific actors, as will be shown below. Moreover, computer simulations are ascribed a new level of manipulativeness (or "*Gefügigkeit*") (Merz 2003, p. 273). On the flipside, simulations show unprecedented complexity. In this regard, Sismondo (1999) speaks of simulating as

---

[3]Sismondo has defined "[s]imple models and complex simulations [...as] endpoints of a continuum. [...] Complex computer simulations can be said to use models of many types" (Sismondo 1999, p. 253).

experimentation, in which the main subject of investigation is the simulation behavior itself. Against this background, there is a lively debate both within and beyond the boundaries of science about how far computer simulations can and should surrogate "real-world" experiments. Another perspective deals with the question of how far simulations can be considered as just another form of empirical data next to others (e.g., Edwards 1999; Morgan 2005). Concerns in regard to the epistemic status of computer simulations have been expressed, for instance, by so-called climate sceptics (see, e.g., Oreskes and Conway 2010). Sceptics have enforced the difference between "real data" and "mere simulation" and urged policy-makers to wait for "sound evidence" before setting counter-measures to combat anthropogenic climate change. Thus, computer simulations find themselves in the middle of credibility battles. Against the backdrop of their highly contested nature, it appears especially interesting whether and how computer simulations and models can become the "currency" of all the different heterogeneous actors involved in transdisciplinary research, in the sense that the results gained by means of computer simulations are accepted by the respective peer communities.

Looking at the projects under investigation, one can see that different kinds of simulations and models are being developed and deployed. What all of them have in common is that they are at some point dealing with the management of the relations between nature and culture. Some of them are concerned with the study of the national, local, and/or regional dimensions of global change. The main goal is therefore to gain insights into the compatibility of local socio-economic development and ecological principals. The applications of computer simulations thus deal, for instance, with the question as to how the demands of several stakeholders can be met while not jeopardizing biodiversity within a distinct region. A key feature of (computer) simulations is the role of a scenario. An example for such a scenario would be that in the near future Austria produces all the food its people need within the country. Based on this assumption, questions—such as "Can Austria provide all the food to guarantee a healthy nutrition of its inhabitants?"—are posed. In doing so, experiments can be probed without—and that is also one of the main arguments for the use of simulations in general—material intervention and real life consequences. As will be discussed below, scenarios bind different disciplinary sub-models together. Furthermore, based on simulations, it seems possible to build a frame in which disciplines can move—in the very sense of a boundary object (Star 2010, see below).

4 Materials and Methods

I have investigated the Austrian research program proVISION[4] in the area of sustainability research and the projects funded by it within the last decade, answering basically the research question of how an integration of both scientific and extra-scientific knowledge can be realized. The program has strongly encouraged attempts to bring various kinds of expertise and knowledge together. This is considered as a chance to handle complex societal problems we are facing alongside the society-ecology-economy triangle. In line with concepts of new modes of knowledge production (see above), special emphasis is put on the integration of extra-scientific actors in knowledge production processes as well as the applicability of the knowledge produced in the context of a certain societal problem.

To carve out what integration of knowledge stemming from heterogeneous scientific and extra-scientific actors means in terms of concrete practices, I can utilize a wide range of materials collected in the research project "Transdisciplinarity as Culture and Practice".[5] Interested in the micro-processes of collaborative involvement in transdisciplinary research, the methodological work identified eleven projects, which were analyzed in depth. A total of 29 interviews were conducted with various scientific and extra-scientific actors, and eleven project meetings of three ongoing projects were observed. Moreover, one focus group was conducted with scientists interested in reflecting on the reality of transdisciplinary collaboration, in which additionally available materials such as project reports, publications, and proposals were taken into consideration.

In this paper, I will specifically focus on sites in which computer simulations and models are presented, developed and discussed, such as the program and project outlines on the respective websites, project proposals, publications, and other forms of dissemination. In addition, the interview and focus group materials

---

[4]The project website (www.provision-research.at) is no longer online but still can partly be accessed via WayBackMachine: http://web.archive.org/web/20131212181724/http://www. provision-research.at/ (Accessed 18 May 2015).

[5]The project was carried out by Ulrike Felt (project lead), Andrea Schikowitz, Thomas Völker, Dorothea Born, and me. See http://sciencestudies.univie.ac.at/forschung/abgeschlossene-projekte/transdisciplinarity-as-culture-and-practice/ (Accessed 5 Sep 2014).

as well as the protocols of the observed project meetings enrich and supplement the insights gained from the analysis of documents, and support the development of a comprehensive picture of the use of models and simulations in transdisciplinary research.[6]

The analysis of the data is oriented toward a newer method of grounded theory developed by Clarke (2005), called "situational analysis". Various elements to be found within the situation of inquiry as well as the (dis-)entanglements among them lie in the center of analytical attention. This allows us to grasp how the specific situation of crafting computer simulations for the integration of heterogeneous actors is shaped by the "[…] individual, collective, organizational, institutional, temporal, geographical, material, discursive, cultural, symbolical, visual and historical conditions" (Clarke 2005, p. xxii).

5   Empirical Findings

In times in which a considerable amount of research is organized in projects, it is becoming ever more important not only to look at the projects themselves but also at the programs that are funding them and at the requirements and prescriptions they set. Accordingly, I start by outlining how the concrete research program imagines collaborative participation in transdisciplinary sustainability research. Afterwards, I present the insights gained from the analysis of the projects as funded by the program.

---

[6]For reasons of anonymization, I will refer to interview numbers and the line in which the interviewee expressed a sentence or phrase (e.g., "interview scientists 1, 345"). In cases I quote from proposals, reports, publications, protocols, etc. of one of the investigated projects, I will tag that as, e.g., "proposal project x", or "publication project y". The interview quotes are originally in German. I translated all the quotes into English, the same holds for some of the quotes taken from other materials, such as proposals, presentations at homepages, etc. Different kinds of models and simulations can be found in the investigated transdisciplinary research project. For sure, certain arguments, ways of legitimization and description will hold for one kind of modeling and simulating more than for another. Basically, however, my results echo characteristics and dimensions that hold for all computer simulations and models and represent the dominating ways of arguing and reasoning for their deployment in transdisciplinary sustainability research.

5.1 Vague Prescriptions and Powerful Moments: How the Research Program Does (Not) Prescribe Transdisciplinary Knowledge Production

The research program defines "transdisciplinarity" as "[...] the sort of scientific work in which partners from outside the scientific community actively take part within the production of knowledge" (Begusch-Pfefferkorn 2005, p. 5).[7] Designed along the lines of "mode 2 knowledge production" (Gibbons et al. 1994), people rooted in different scientific and extra-scientific communities and institutions are encouraged "[...] to come together in temporary working teams and networks which dissolve when a problem is solved or redefined" (Gibbons et al. 1994, p. 6).

Notably, the prescriptions concerning the concrete steps towards integration within transdisciplinary knowledge production remain very vaguely formulated by the program. At some point this is not surprising, as it seems to be the very point of transdisciplinary research to be vague—everything and everyone is supposed to flexibly assemble and reassemble around a certain problem, as is indicated in the following quote:

> As a rule, actors with extremely divergent perspectives, interests, and work styles thereby encounter each other. In contrast to the established routines and methods of work in the framework of scientific disciplines, in such heterogeneous contexts, practical research work must be learned a new each time, because the problems arising in every life always demand their own specific form of a research setting (Bergmann et al. 2005, p. 9).

It thus seems to be the very idea of transdisciplinary knowledge production not to produce knowledge according to pre-described standards, not to have pre-set rules.

Also, the vagueness should not be an entry point for critique. Akrich (1992) and Suchman (2007) have highlighted in different contexts that plans or scripts can never anticipate all the conditions under which they will be played out in "real-world" situations. There will always be unforeseeable circumstances, resistance and workarounds. However, this vagueness leaves decisive power to the project collaborators, so they can select adequate integration procedures, instruments, and equipment for the creation of "socially robust knowledge" (see above). In the context of the GMO controversies, Böschen (2009) describes a situation in which heterogeneous actors try to find consensus as a confusing situation, which he

---

[7] Original German text: "In proVISION steht Transdisziplinarität für jene wissenschaftliche Arbeit, in der außerwissenschaftliche Partner, Partnerinnen an der Entstehung des Wissens mitwirken" (Begusch-Pfefferkorn 2005, p. 5).

defines as an "epistemic no man's land" (p. 510). He emphasizes that—while usually not addressed in the political debates—the arguments offered by different actors need to be understood as "[...] strategies to frame the scope of conflict" (p. 517), as sources of power. Following this line of argument, the moments in which certain research instruments or methods are developed and presented can be considered as key moments in which orders are created, hierarchies among actors produced and reproduced, power relations negotiated, roles and responsibilities redistributed, and—in the broadest sense—the relations between science and society reconfigured.

While the concrete ways in which transdisciplinary knowledge is produced are hardly touched by the research program, a broad consensus has been reached about what the adequate means for producing knowledge capable of integrating heterogeneous kinds of knowledge and expertise is, namely computer modeling and simulation. In the following, I describe how the computer simulations and models are discursively designed in the analyzed transdisciplinary projects.

5.2   Arguing for the Use of Computer Models: The Dream of an Overall Model

When looking at the arguments offered in regard to the attempt to make use of computer simulations and models for the participatory production of knowledge in project proposals and in other dissemination formats (e.g., project reports and publications), one major narrative ascribes to computer simulations the ability to bring different kinds of knowledge together. The need to do so is derived from the idea that knowledge produced within single disciplines and interest areas is deficient, while the construction of computer models and simulations is expected to support the production of a more holistic view, a big picture. It says for instance:

> The idea: Different knowledge bricks are assembled to a whole. Knowledge about hydrology, ecology, political economy and history can be reconciled (presentation@program homepage, project 6).

> The inter- and transdisciplinary overall image (German orig.: Gesamtbild) originates from the collaboration of natural, economic, social and human scientists, the population, schools, tourist and cultural associations, and economic and political decision-makers on site (presentation@program homepage, project 11).

To assemble more holistic knowledge from different knowledge bricks, a "*Gesamtmodell*"—as it is called in one project proposal (project 7)—is crafted. The term *Gesamtmodell* could be translated as "overall model" or as an "integrated model", a phrase used in another project's publication (project 10). The development of such a model is basically aimed at responding to two gaps in current sustainability research.

First, the attempt is related to the impression that the social element is usually underrepresented in many models that deal with ecological issues. Moreover, natural and social processes are often investigated separately from each other. In response to that, the models developed and used within the projects are considered to be able to contribute to a better understanding of the dynamics between ecological processes and "human" behavior. One project aims to "[...] design 'tools' that do not exclude the human factor, but rather actively include it" (proposal, project 6). To address this scientific challenge (proposal, project 10), existing models are expanded and adapted (proposal, project 7). Data and models from the social, economic, and natural sciences are assembled and related to each other. One project proposal says:

> [The model] is expected to yield considerable progress in the integration of socio-economic and natural-science approaches in [x]-research (proposal, project 10).

Second, a praxis problem (proposal, project 10) is described, which should be made manageable by using models and simulations. The overall model is thus attributed a leading role towards achieving a (more) sustainable management of a specific "problem area", which may be either Austria or a smaller region within it. It comes to be considered a "strategic planning instrument" to be used as the basis for decision-making (proposal, project 7). The insights gained from simulations should form the basis for directives for handling specific problems in certain societal problem areas, as is made explicit in the following quote:

> The aim is to provide practitioners with decision support tools for developing resilient land uses and adaptable social and economic structures (homepage, project 9).

Specifically with regard to the "praxis problem", the demonstrative character of models and simulations is highlighted, as is clear in these passages from the proposal from project 8: "The integration of the spatial view with models for [x]-dynamics can depict change in [y. ...] Different development paths for [z] become conceivable." The visualizations (or "perspectives") produced in the act of modelling and simulating are expected to "[...] help communicate [...]," or "[...] to overcome barriers of different languages [...]," and allow us to "[...] think across management perspectives and different societal accounts as well as scientific accounts from different disciplines." In this sense, models and simulations are framed as a kind of translation instrument.

The aim to handle two challenges (a scientific one and a praxis problem) within the construction of one overall model is related to Keller's (2000) analysis of computer models and their quality of simultaneously being both models "for" and models "of". The act of establishing a comprehensive understanding of the relationship between nature and society is already connected to the development of a "future managing instrument" expected to be useful for "strategic intervention" in a certain societal problem context (report proposal, project 7).

When it comes to the ability to integrate knowledge that has been produced by different people in different times and places, computer simulations are often portrayed as democratic tools, which support the balanced integration of different perspectives:

> Natural and social sciences contribute equally to the creation of predictive models with respect to large-scaled succession patterns in the course of potential [x-] management alternatives (publication, project 6).

> The specificity is that simply a computer model is being developed that different disciplines can access to an equal extent and that different disciplines must really be able to use as model (interview scientist 6, 1193).

The quotes exemplify how closely the development of a computer model is interwoven with the idea of bringing heterogeneous kinds of knowledge together and to do it in a way in which no sort of knowledge is given preference to others. In view of multitude of collection, alignment, and harmonization activities, computer simulations and models are also considered as research objects themselves.

> It is also a long process of learning. It is actually […] a huge model […]. When one turns 90, one still won't know everything about the model (interview scientist 18, 532).

The quote shows that the models and simulations become research objects themselves in the sense that the behavior of the model itself ought to be understood by its constructors.

5.3 Simulating and Modeling as Balancing Act Between a Multiplicity of Attributions

The analysis of moments in which use and development of computer simulations and models are negotiated makes palpable the hybrid nature of models and simulations, as highlighted in former studies (see Sect. 3). While the aim of the projects is to simulate certain developments, and in doing so to understand interrelations and dynamics, such as the interrelation between human actions and biodiversity, the same simulations are expected to be useful for strategic intervention. The following quote makes this point:

> Beside well-founded results about the potential for development of the [x-]area, the project laid the basis for a collective planning process of the responsible decision-makers, in which the developed models can be incorporated (final report, project 6).

A multiplicity of attributions ascribed to computer simulations (research object, instrument to understand the relations between humans and nature, a strategic planning instrument) is making them hybrids being and doing many different things at once. This hybridity plays out in an interesting way when the construction processes of simulations within project meetings is considered, namely as a "trade-off." Concretely, the trade-off is circling around the following concerns:

- What can be accomplished within the framework of a research project? The computer simulations carried out always mirror the available data and the modeling expertise of the participants. In addition, the time frame of the projects limits the scope of the computer simulations performed (e.g., in terms of dimensions considered or in terms of scenarios developed). "This clearly exceeds the means we are provided in the project" (meeting, project 1), is one of the phrases used when it comes to legitimize the reduction of complexity in the process of model development.
- What are (un-)likely and (un-)desired futures? Based on the state of the art in the respective scientific and extra-scientific communities, and common ideas about (un-)likely futures, trends are identified from which scenarios are derived. These are also related to (un-)desired futures of (imagined) users of knowledge to be produced by the simulations.
- What can be shown and approved? Decisions are also made with regard to the respective audiences to be convinced and their (cognitive) preferences. Typically, a time span is chosen from which it can be expected that "[...] one can really see the difference" (meeting, project 2), as one scientific project collaborator stated in a project meeting, pleading for the deployment of "extreme scenarios" due to their "show effect" (meeting, project 2).

Producing models and simulations in transdisciplinary research becomes a balancing act between what the scientists aim to achieve in regard to the state of the art in their discipline, what can be managed within the frame of the project, and ideas about the ways the produced data are used and read. All these kinds of reasoning inform the decision-making processes within the construction of models and simulations.

5.4 Involvement of Scientists and Extra-Scientific Actors in the Development of Computer Models and Simulations

During our studies we observed a considerable number of project meetings in different constellations, in which the construction of the overall model was in the center of attention. In some cases, only the scientific core teams were invited, while in others *practitioners* and other extra-scientific actors (administrators, teachers, NGO representatives, etc.) were present. In either case, the schedule was characterized by separate presentations of preliminary results by largely scientific sub-teams, followed by discussions in which other participants posed questions. This meeting's structure itself hints to the fact that disciplinary and institutional boundaries do not necessarily need to be blurred in so far as the specific expertise within a certain scientific area is not questioned or challenged by the other participants, at least not when it comes to the established ways of manipulating data. A project leader makes this "keeping untouched" explicit:

> There needs to be collective cooperative work, but each one needs to bring things from his/her discipline, things which are specific to this field (interview scientist 2, 568).

Another scientist states:

> It is obvious that everybody uses his/her ideal methods to achieve the objectives (interview scientist 17, 1023).

In this sense, computer models and simulations work as "boundary objects" (Star and Griesemer 1989) in their ability to do both, binding different perspectives together but also keeping them apart, as they allow the coexistence of diverging approaches and methods without further complications. Not all participants seem to find entrance into this kind of knowledge production equally easily, however. A project participant describes how some disciplines are simply too different from the others to be involved into modeling activities:

> Well, methods, that is rather difficult, because each discipline has its own methods, and many methods that make sense in the natural sciences make no sense at all in the social area. [...] Some methods that the discipline x uses, they don't make any sense for us [...] (interview scientist 17, 861).

A non-natural scientist describes how she tried to transform her knowledge into a "model-able" one:

> With the [x], I tried it once. And then I got such a cluster from [Sibille]—such a [x] cluster—which is based upon [Andrea's] data and I have—well [Sibille] did that—superimposed [x] with [y]. So we have a map now, from which I cannot really read out anything now, of course. I can say, okay it is blue—but that has to be done by an expert, but I think one could use that (interview scientist 20, 1299).

That the ideal methods are kept untouched thus only holds for those participants who can provide the sort of aggregated knowledge that simulations and models accept and are able to process. And it seems that scientists who do not model are shifted from the core of knowledge production to the science-society borderland, where they take over responsibility for translating knowledge to diverse audiences, for instance.

Moreover, a separation is created between those who are calculating and integrating data, for whom the models are research instruments—namely the core scientists—and those for whom the models are "future-managing instruments" (see above)—i.e., the extra-scientific actors. Accordingly, scientific and extra-scientific actors assume different roles regarding the construction and deployment of models and simulations. In line with this, extra-scientific actors are often considered as "users" from the beginning of the knowledge production processes and as such not directly "touching" the model. One of the authors puts it as follows in an interview: "[…t]he development of an interdisciplinary model within a transdisciplinary process" (interview scientist 6, 1208). By using qualitative methods such as focus groups, workshops, or interviews, the data of local actors are gathered and then translated into data that can be modelled (e.g., interview scientist 6, 1203). The scientists function as intermediaries between the model and extra-scientific actors. The transdisciplinary process of the development of the interdisciplinary model remains strongly focused on customization and on the users' wants and needs. The customization process can involve different testing phases, in which users are concerned with the improvement of the simulation's user interface:

> Interface elements (such as sliders and graphs) can be easily implemented allowing users such as regional stakeholders to test model versions at all stages of model development (publication, project 10).

In this context, a model is ascribed a certain immediateness, which motivates extra-scientific actors to engage with it. The users are allowed to witness "in real time" how their decisions will make a difference in their future: "[L]etting these data flow into a model that is then put in operation before their eyes" (publication, project 10). Extra-scientific actors thus become "pre-inscription witnesses" (Akrich 1992), which is in turn expected to increase both "learning processes and empowerment of the local population" (publication, project 10).

The empirical findings have shown that the development of computer models and the use of computer simulations in transdisciplinary sustainability research are intimately related to the ability to integrate a variety of knowledge stemming from different backgrounds. This approach produces knowledge that can be used as a basis for decision-making in certain societal contexts. In light of the multiple functions ascribed to computer simulations and models by diverse participants, their configuration becomes an act of balancing and translating. In the course of their configuration not only the models and simulations are developed but also the participating actors themselves are "constructed" as are their roles and responsibilities in transdisciplinary research. The notion, as developed in the following, is aimed to reflect on the dynamics between calls for collaboration between heterogeneous actors and the deployment of computer models and simulations to make such collaboration possible.

6 Computer Models and Simulations as "Integration Machines"

Inspired by the ways that models and simulations are characterized in the transdisciplinary sustainability projects that I have investigated, I will now develop the notion of the integration machine, bringing together the empirical insights with theoretical concepts in order to establish a basis from which I can reflect more broadly upon the rise of computer modeling and simulations in a changing landscape of knowledge production, in which collaboration between diverse scientists and extra-scientific actors becomes increasingly common.

While models and simulations have to be understood as heterogeneous entities on many levels, a joint characteristic seems to be the description as instruments for bringing together the knowledge and expertise stemming from a variety of actors. My intention, however, is to go beyond characterizations of computer models and simulations as integrative tools or instruments. The notion of integration machines is intended to capture the dynamics of calls for integration and the epistemic practices deployed to achieve such integration. Accordingly, I define "integration machines" as socio-material assemblages being built and stabilized around computer modeling and simulating in transdisciplinary sustainability research. In doing so, I follow a tradition in science and technology studies that seeks to understand scientific and technological developments in terms of the practical forms of aligning meaning and matter. Latour (1987) highlighted the collective efforts of translation between (human and non-human) actors within the stabilization of networks. Knorr Cetina (1999) also did so, for instance, when paying attention to the establishment of specific arrangements of resources which in a given field, make up how we know what we know.

Moreover, Deleuze's and Guattari's (1977a) conception of machines, or more specifically of human-machine relations, is constitutive for my understanding of integration machines. They state:

> The object is no longer to compare man and the machine in order to evaluate the correspondences, the extensions, the possible or impossible substitutions of one for the other, but to bring them into communication to show how [hu]man is a component part of the machine, or combines with something else to constitute a machine (Deleuze and Guattari 1977b, p. 125).

Following this definition, we can no longer conceptualize computer simulations as tools or instruments. They rather become alignments of different actors (human and non-human), which together constitute a machine. In doing so, they reveal productive forces. They re-produce and stabilize identities and ways of being within the science-society relationship, and they perform realities.

The following sections are devoted to the human and non-human actors that assemble to form integration machines, as well as to their performative dimensions.

6.1 How Integration Machines Are Made of Ideas, Imaginations, Hopes, and Promises

Significantly, dramatic and dystopian narratives about a new dimension of societal challenges are connected to the need to deploy computer simulations. As introduced above, we witness an increase in the awareness of the scope of power that can and must be granted to humans with respect to their environment. The imperative to accept responsibility for the environment we are (co-)creating translates into the idea that the society-economy-ecology triangle needs to be managed (see Introduction, Sect. 1). And it is exactly the ascribed forward-looking while not (yet) "materially intervening" character (see, e.g., Morgan 2003) that seems to make computer simulation and modeling the appropriate means to cope with these newly accepted challenges.

In consideration of the ever increasing computational power, we additionally witness a renaissance of the hope that we will finally be able to craft a big picture, or as Edwards puts it in the context of climate science, a "consistent global data image" (Edwards 2010, p. 420), from which political measures can be derived. Incorporating huge amounts of knowledge and data produced by different people in different times and places, the consistent data image promises to provide us with insights so urgently needed for coping with the problems of immense complexity we currently face.

These examples show how "integration machines" are tied to "a broader technoscientific economy of promises" (Felt and Fochler 2010, p. 18), how imaginative resources take part in the creation of the integration machine.

6.2 How Integration Machines Are Black Boxing, Delegating, Integrating, and Sorting Out

We have come to know computer simulations as hybrids—being and doing many different things at once. While one of their aims is to simulate certain developments and in doing so to understand interrelationships and dynamics, the very same simulations are expected to be useful to determine strategic intervention. Moreover, they are introduced as translation instruments with a demonstrative side that supports learning processes (see Sect. 5).

How the multiplicity of attributes is balanced in terms of assumptions and data that inform the construction of models and simulations is often not apparent to those who are interested in assessing the results or in re-using the data and models produced. In this sense, integration machines are "black boxing" their conditions of production, while evoking the impression of mechanical objectivity (Daston and Galison 2007). As Silvio Funtowicz and Jerome Ravetz emphasize in this regard:

> Computer models are the most widely used method for producing statements about the future based on data of the past and present. For many there is still a magical quality about computers, since they are believed to perform reasoning operations faultlessly and rapidly (Funtowicz and Ravetz 1993, p. 742).

Computer simulations, however, can be rather characterized as a composition of collected, re-used, and adapted models and data. Before a simulation can be run, there is enormous complexity and uncertainty that needs to be coped with and reduced. Furthermore, different databases need to be aligned to each other, or "harmonized" as one of the project participants calls it (meeting, project 1), which is another source of friction and transformation:

> Every interface between groups and organizations, as well as between machines, represents a point of resistance where data can be garbled, misinterpreted, or lost (Edwards et al. 2011, p. 3).

Shipworth (2013) has pointed to a related difficulty. Developing the funny-sounding notion of the "epistemic sausage machine", he alludes to a specific way energy models are functioning as black boxes:

> The outputs of the model carry no information regarding the forms of knowledge, their basis or intention of which they are constituted. This makes building energy models akin to some kind of epistemic 'sausage machines'—combining inputs of all qualities and types into outputs of homogenous and indeterminate quality and type (Shipworth 2013, p. 254).

Computer simulations are deployed to produce knowledge in transdisciplinary research because of their ability to integrate knowledge and expertise stemming from a variety of actors. The analysis, however, has shown that this holds only for the sort of aggregated knowledge that simulations and models can accept and are able to process. In this sense, integration machines are not only integrating but also "sorting out" (Bowker and Leigh Star 1999). This dynamic seems especially critical with respect to the question as to which knowledge will be kept and which knowledge will get lost in the long run.

As has been said elsewhere, the notion of the integration machine is inspired by the way computer simulations and models are portrayed within the transdisciplinary sustainability projects themselves. In publications, proposals and reports they are often characterized as a sort of democratic tool, which partly takes over the responsibility of integrating knowledge stemming from different scientific disciplines and extra-scientific actors. This portrayal provokes the idea that the processes of integration are delegated to the computer simulations for their ability to automatically choose and select knowledge in a symmetric and democratic way. The integration processes, however, can be described in terms of translation and transformation (Akrich and Latour 1992) rather than as automation, as has also been emphasized in a different context:

> In contrast to any naïve view of automation, delegation to artefacts results not in the wholesale transfer of human work across a sociotechnical divide, but rather a reconfiguration of that work bringing new technological and human actors into the mix (Ribes et al. 2013, p. 10).

How exactly different participating actors are brought into the mix in the course of the design of integration machines is discussed in the next section.

6.3 How Integration Machines (Re-)Produce Ways of Being in the Science-Society Relationship

When Lucy Suchman pays attention to human-machine reconfigurations she addresses "questions of difference—and more particularly asymmetries [between humans and machines]" (Suchman 2007, p. 267). With regard to integration machines, in contrast, it seems especially valuable to deal with asymmetries between the different human actors participating in transdisciplinary research and becoming parts of integration machines.

We encountered (natural) scientists mostly as producers and designers of models and simulations, and extra-scientific actors often as users who add specific concerns and interests or a specific problem view. Distinct identities, responsibilities, and authorities are thus very well kept. This seems to be connected to well-established dichotomies—such as "emotion- and interest-driven" societal actors versus "rational," "strategic," and "objective" scientific experts—that are important resources for public trust in science (see, e.g., Haraway 1997). Models and simulations seem to offer the possibility to uphold a "snow white fairy tale of scientific objectivity" (Farrell 2011, p. 354), and maintain a certain autonomy, which guarantees the acceptance of the produced knowledge in the respective communities. While "integration machines" allow us to keep certain boundaries and heterogeneities between different actors (such as the differentiation between knowledge producers/designers, knowledge translators, and knowledge users), at the same time they seem to put great pressure on those who have never been involved in modeling and simulation. As in global climate science, in transdisciplinary sustainability research "integrated models [seem to] serve a central organizing function for large and growing epistemic communities, both within and beyond science" (Edwards 2010, p. 421). This means that integration machines seem to be simultaneously doing both integrating and disintegrating knowledge, overcoming boundaries while (re-)producing others.

6.4 How Integration Machines Produce Realities

Last but not least, integration machines also have a strong prescriptive dimension. Certainly, they are fictional in character and clearly not materially intervening in the sense that a certain region would change just because simulations were performed. Nevertheless, integration machines are producing realities. What Deleuze (1989) has conceptualized as a "time-image" in his analysis of movies, seems to hold for the simulation too: it co-produces the virtual and the actual. Subsequently, it may be valuable not to think of simulations and models as machines that are producing virtual realities which stay separate from our "real" realities, but of ones that are connecting the actual and the virtual, and in doing so (re-)produce our past(s), present(s), and future(s). To give an example, the simulation of futures is sometimes connected to a deficient or problematic regional development in the past. As an integral part of simulations, such narrations are crucially prescribing what is to be considered as deficient or problematic or what as a desirable or less desirable development.

Moreover, a range of presumptions inscribed into computer simulations remains unquestioned. This holds for the imperative to assume responsibility for present and future nature-culture relations as well as for the trust in numbers (Porter 1995). Against this background, one could speak of "anti-politics machines" in a similar way as Dourish (2010) has in the context of the design of new information technologies for the promotion of environmental sustainability. He highlights that

[p]olitical, social, cultural, economic, and historical contexts have critical roles to play, not only because they shape our experience with information technologies, but also, and even more, because information technologies in contemporary life are sites at which these contexts are themselves developing (Dourish 2010, p. 8).

Integration machines seem to operate as anti-politics machines as they are producing realities, connecting pasts to desirable and less desirable futures, and prescribing problems as well as the ways to solve them, while not leaving much room to question the actual ways of producing realities, of constructing problems or of reflecting on the ways in which actors are assembled to integration machines.

7 Conclusions

Against the background of complex and pressing socio-environmental problems such as climate change—which seem to outweigh the capacities of single scientific disciplines—, alternative modes of knowledge production are currently being probed. A central feature of alternative modes of knowledge production is the call for the integration of heterogeneous—also extra-scientific—actors into knowledge production processes. I began intending to investigate how these calls for integration are translated into integrative research practices in different transdisciplinary projects as funded by the Austrian research program proVISION in the area of sustainability research, which were identified as prototypes of new modes of knowledge production. Based on the observation that most of the analyzed projects strongly relied on computer simulation and modeling in order to integrate heterogeneous kinds of knowledge stemming from scientific and extra-scientific actors, I developed the notion of the integration machine. The aim of this notion is to grasp the dynamics of attempts to integrate heterogeneous and extra-scientific actors into the processes of knowledge production and the specific epistemic practices employed to do so. It allows us to pay attention to alignments of material and symbolic recourses and of human and non-human actors that are being stabilized around the deployment of computer simulations.

Other terms have been coined to understand collaboration processes between heterogeneous (scientific) actors—such as the notion of the "boundary object" (Star and Griesemer 1989) or the "trading zone" (Galison 1996). They often focus on how collaboration and coordination between heterogeneous actors are made possible. The notion of the integration machine specifically attempts to attend to the productive dimensions, such as the re-production of orders and re-distribution of responsibilities in science-society relations, the production of realities, or the asymmetries between different actors and kinds of knowledge constituted by machines in the course of attempts to integrate heterogeneous kinds of knowledge.

The aim of the analysis was to show how integration machines are not only integrating, but also excluding knowledge. Substantially, the establishment of computer simulations in the field of transdisciplinary sustainability research seems related to the promise that they would provide a basis for political decision-making. As a consequence, inscribed logics—such as the orientation towards futures, the trust in quantified knowledge, the idea that nature-culture relations need to be responsibly handled by extensive monitoring, or the use of simulations and models themselves—never become part of participatory processes of negotiation. Thus, integration machines also need to be regarded as anti-politics machines (Dourish 2010; see above), and arrangements that are not only opening up put also closing down participatory involvement (Stirling 2008).

The analysis has shown that potential societal uses play an important role in the processes of model development. The specific forms of involvement and expertise that are granted to extra-scientific actors, however, could benefit from further creative rethinking. So far, these seem too closely oriented towards a model of knowledge production in which scientists assume the role of creators and designers of computer simulations, while societal actors are usually conceptualized as users. More finely grained knowledge also seems to be needed with respect to the dynamics of the multiple "forms of being" (e.g., visualizing tool, planning instrument, research object) that "co-inhabit" computer simulations.

Acknowledgments I gratefully acknowledge the Austrian Ministry of Science and Research for funding the research project "Transdisciplinarity as Culture and Practice." I moreover want to thank all the interviewees for sharing their views on transdisciplinary knowledge production and providing insights into transdisciplinary work practices. Special thanks go to Ulrike Felt, Maja Horst, Stephen Hilgartner, Katrin Igelsböck, Christoph Musik, Andrea Schikowitz, Judith Simon, and Thomas Völker (in alphabetical order), and, last but not least, to the editors of the book for providing invaluable feedback to previous versions of the paper.

References

Akrich M (1992) The description of technical objects. In: Bijker WE, Law J (eds) Shaping technology/building society. Studies in sociotechnical change. MIT Press, Cambridge, pp 205–224

Akrich M, Latour B (1992) A summary of a convenient vocabulary for the semiotics of human and nonhuman assemblies. In: Bijker WE, Law J (eds) Shaping technology/building society. Studies in sociotechnical change. MIT Press, Cambridge, pp 259–264

Begusch-Pfefferkorn K (2005) proVISION: Forschungsprinzipien. http://www.provision-research.at/. The homepage is no longer online, but a brief outline of the program can be found at the homepage of the Ministry of Science http://wissenschaft.bmwfw.gv.at/bmwfw/forschung/national/programme-schwerpunkte/provision/. Accessed 18 May 2015

Bergmann M (2003) Indikatoren für eine diskursive Evaluation transdisziplinärer Forschung. Technikfolgenabschätzung. Theorie und Praxis 12(1):65–75

Bergmann M, Brohmann B, Schramm E (2005) Quality criteria of transdisciplinary research. A guide for the formative evaluation of research projects. http://researchgate.net/publication/256437773_Quality_Criteria_of_Transdisciplinary_Research_A_Guide_for_the_Formative_Evaluation_of_Research_Projects. Accessed 4 May 2015

Borup M, Brown N, Konrad K, van Lente H (2006) The sociology of expectations in science and technology. Technol Anal Strat Manag 18(3/4):285–298

Böschen S (2009) Hybrid regimes of knowledge? Challenges for constructing scientific evidence in the context of the GMO-debate. EnvironSci Pollut Res 16(5):508–520

Bowker G, Leigh Star S (1999) Sorting things out. Classification and its consequences. MIT Press, Cambridge

Bunders JF, Broerse JE, Keil F et al (2010) How can transdisciplinary research contribute to knowledge democracy? In: Veld RJ (ed) Knowledge democracy. Consequences for science, politics, and media. Springer, Heidelberg, pp 125–152

Clarke AE (2005) Situational analysis. Grounded theory after the postmodern turn. Sage, Thousand Oaks

Daschkeit A (2007) Zur Beurteilung transdisziplinärer Forschung. Hinweise auf Bücher zu Transdisziplinarität. GAIA 16(1):58–65

Daston L, Galison P (2007) Objectivity. Zone Books, New York

Defila R, Di Giulio A (1999) Transdisziplinarität evaluieren - aber wie? Evaluationskriterien für inter- und transdisziplinäre Forschung. Panorama (Evaluating transdisciplinary research) special issue 1(99):3–27

Deleuze G (1989) Cinema 2: The time-image. University of Minnesota Press, Minneapolis

Deleuze G, Guattari F (1977a) Anti-Ödipus. Kapitalismus und Schizophrenie I, Suhrkamp, Frankfurt

Deleuze G, Guattari F (1977b) Balance sheet—program for desiring-machines. Semiotexte 2 (3):117–135

Dourish P (2010) HCI and environmental sustainability: the politics of design and the design of politics. In: Proceedings of the 8th ACM conference on designing interactive systems. ACM

Edwards PN (1999) Global climate science, uncertainty and politics: data-laden models, model-filtered data. Sci Cult 8(4):437–472

Edwards PN (2010) A vast machine: computer models, climate data, and the politics of global warming. MIT Press, Cambridge

Edwards PN, Mayernik MS, Batcheller A et al (2011) Science friction: data, metadata, and collaboration. Soc Stud Sci 41(5):667–690

Farrell KN (2011) Snow white and the wicked problems of the west: a look at the lines between empirical description and normative prescription. Sci Technol Hum Values 36 (3):334–361

Felt U (2010) Transdisziplinarität als Wissenskultur und Praxis—Transdisciplinarity as culture and practice. GAIA-Ecol Perspect Sci Soc 19(1):75–77

Felt U, Fochler M (2010) Machineries for making publics: inscribing and describing publics in public engagement. Minerva 48(3):219–238

Funtowicz S (2001) Peer review and quality control. International encyclopaedia of the social and behavioural sciences. Elsevier, Oxford, pp 11179–11183

Funtowicz S, Ravetz J (1993) Science for the post-normal age. Futures 25(7):739–757

Galison P (1996) Computer simulations and the trading zone. In: Galison P, Stump DJ (eds) The disunity of science: boundaries, contexts, and power. Stanford University Press, Stanford, pp 119–157

Galison P (1997) Image and logic: a material culture of microphysics. University of Chicago Press, Chicago

Gibbons M, Limoges C, Nowotny H et al (1994) New production of knowledge: dynamics of science and research in contemporary societies. SAGE, London

Gieryn T (1983) Boundary work and the demarcation of science from non-science: strains and interests in professional ideologies of scientists. Amer Sociol Rev 48(December):781–795

Haraway D (1997) Modest_Witness@Second_Millennium.FemaleMan_Meets_Onco-Mouse: Feminism and technoscience. Psychology Press, New York

Hessels LK, van Lente H (2008) Re-thinking new knowledge production: a literature review and a research agenda. Res Policy 37(2008):740–760

Jahn T, Bergmann M, Keil F (2012) Transdisciplinarity: between mainstreaming and marginalization. Ecol Econ 79:1–10

Keller EF (2000) Models of and models for: theory and practice in contemporary biology. Philosophy of Science 67, Supplement. Proceedings of the 1998 Biennial Meetings of the Philosophy of Science Association. Part II: Symposia Papers, pp S72–S86

Knorr Cetina K (1999) Epistemic cultures. How the sciences make knowledge. Harvard University Press, Cambridge

Latour B (1987) Science in action. How to follow scientists and engineers through society. Harvard University Press, Cambridge

Latour B (2011) Waiting for Gaia. Composing the common world through arts and politics. A lecture at the French Institute, London

Le Dantec CA, DiSalvo C (2013) Infrastructuring and the formation of publics in participatory design. Soc Stud Sci 43(2):241–264

Lund Declaration (2009) Europe must focus on the grand challenges of our time. Swedish EU Presidency. 8. July 2009, Lund, Sweden

Max-Neef M (2005) Foundations of transdisciplinarity. Ecol Econ 53(1):5–16

Merz M (2003) Die Simulative Ordnung der Dinge. In: Michel M (ed) Fakt & Fiktion 7.0. Zwischen Wissenschaft und Welterzählung: Die narrative Ordnung der Dinge. Chronos, Zürich, pp 271–273

Merz M, Hinterwaldner I (2012) Neue Bilder, Modelle und Simulationen: Zwischen Repräsentativität und Produktivität. Handbuch Wissenschaftssoziologie. Springer Fachmedien, Wiesbaden, pp 303–316

Morgan MS (2003) Experiments without material intervention: model experiments, virtual experiments and virtually experiments. University of Pittsburgh Press, Pittsburgh

Morgan MS (2005) Experiments versus models: new phenomena, inference and surprise. J Econ Methodol 12(2):317–329

Nowotny H (2003) Democratising expertise and socially robust knowledge. Sci Public Policy 30(3):151–156

Oreskes N, Conway EM (2010) Merchants of doubt: how a handful of scientists obscured the truth on issues from tobacco smoke to global warming. Bloomsbury Publishing, London

Pohl C, Hirsch Hadorn G (2008) Core terms in transdisciplinary research. In: Hirsch Hadorn G et al (eds) Handbook of transdisciplinary research. Springer, Dordrecht, pp 427–432

Porter TM (1995) Trust in numbers. The pursuit of objectivity in science and public life. Princeton University Press, Princeton

Ribes D, Jackson SJ, Geiger S et al (2013) Artifacts that organize: delegation in the distributed organization. Inf Organ 23(1):1–14

Shapin S (2012) The ivory tower: the history of a figure of speech and its cultural uses. Br J Hist Sci 45(01):1–27

Shipworth D (2013) The vernacular architecture of household energy models. Perspect Sci 21(2):250–266

Sismondo S (1999) Models, simulations, and their objects. Sci Context 12:247–260

Star SL (2010) This is not a boundary object: reflections on the origin of a concept. Sci Technol Hum Values 35(5):601–617

Star SL, Griesemer JR (1989) Institutional ecology, "translations" and boundary objects: amateurs and professionals in Berkeley's Museum of Vertebrate Zoology, 1907–39. Soc Stud Sci 19:387–420

Steffen W, Grinevald J, Crutzen P et al (2011) The Anthropocene: conceptual and historical perspectives. Philos Trans R Soc A: Math Phys Eng Sci 369(1938):842–867

Stirling A (2008) "Opening up" and "closing down": power, participation, and pluralism in the social appraisal of technology. Sci Technol Hum Values 33(2):262–294

Suchman L (2007) Human-machine reconfigurations: plans and situated actions. Cambridge University Press, Cambridge

Weingart P (ed) (2008) How robust is "socially robust knowledge"? Knowledge and democracy: a 21st century perspective. Transaction, New Jersey

Cosmology and the End of Weberian Science

Genco Guralp

Abstract

The 2011 Nobel Prize in physics was awarded to two teams which, working independently, confirmed the striking fact that the expansion of the universe is accelerating. For many cosmologists, this prize marked another major point in the chain of successful results cosmology obtained in its relatively short history of being an experimental science. In fact, modern cosmology prides itself for becoming a "precision science," breaking sharply with its "speculative" past. I analyze this experimental turn in cosmology and examine different forms of interdisciplinary transgressions that this epistemic shift is built on. I propose that these transgressions that cosmology engenders in the process of establishing its scientific legitimacy attests to the fact that a crucial aspect of the way scientific knowledge is usually characterized is being challenged today. This characterization, which found, as I argue, one of its best conceptualizations in Weber's classic *Wissenschaft als Beruf*, is summarized in his famous "disenchantment" thesis proposing that there exists a sharp boundary between the questions under the jurisdiction of science and the questions of "meaning" (such as theology), which science refuses to answer. I argue that, as the current practice of cosmology confronts this boundary, the assumptions pertinent to the social and epistemic contexts within which scientific knowledge comes into existence will also be put into question, which is what we witness in the case of modern cosmology.

G. Guralp (✉)
Baltimore, USA
e-mail: genco.guralp@gmail.com

© Springer Fachmedien Wiesbaden GmbH 2016
B.-J. Krings et al. (eds.), *Scientific Knowledge and the Transgression of Boundaries,* Technikzukünfte, Wissenschaft und Gesellschaft / Futures of Technology, Science and Society, DOI 10.1007/978-3-658-14449-4_7

Keywords

Cosmology · Interdisciplinarity · Boundary transgressions · Weberian science ·
Methodology

1 Introduction

The 2011 Nobel Prize in physics was awarded to two teams which, working
independently, confirmed the striking fact that the expansion of the universe is
accelerating. For many cosmologists, this prize marks another major point in the
chain of successful results cosmology has obtained in its relatively short history of
being an "experimental science." In fact, modern cosmology prides itself for
becoming a "precision science," breaking sharply with its "speculative" past.

This paper aims to probe this transformation that cosmology underwent, from
being a speculation on the "heavens," to a legitimate empirical science: employing
data-driven computational models, building big science detectors like the Hubble
Space Telescope, and providing mathematically robust explanations of the large
scale structure of galaxies. I study this transformation on the basis of the relationship between the multidisciplinary constitution of cosmology (to be explicated
below) and the picture of scientific knowledge that this constitution presents us
with. I will be particularly attentive to the way this transformation and constitution
is represented within the cosmology literature itself: the *auto-representation* of
cosmology. Through a historical analysis of several key texts, I will claim that this
transformation process is very closely tied to various forms of boundary transgressions in and across disciplines, and I question the roles they perform.[1]

---

[1]There is one important question that needs to be clarified for the purposes of this paper,
concerning the relationship between boundary transgressions and interdisciplinarity. For one
can ask whether each interdisciplinary formation corresponds to a boundary transgression,
or, how a boundary transgression is different from a cooperation between two disciplines. To
answer this question, we need to clarify what we mean by a "transgression": When does a
discipline transgress its boundary? Although it is beyond the scope of this paper to provide a
general treatment, I propose that in any interdisciplinary formation there is an irreducible
element of transgression for the following reason: When a given discipline is employed to
address questions and problems that previously were classified as belonging to another one,
it will need to adjust those questions to its own language and "picture of reality", in order to
be able to deal with them (here, one good example could be the situation in molecular
genetics and macroscopic biology). I suggest that it is appropriate to understand this form of
disciplinary adjustment as a transgression. To be sure, I do not claim that this is the only

To this effect, I distinguish three levels of transgression. On what one may call the lowest level, there exist transgressions within a given traditional discipline, or *intra-disciplinary* transgressions. Thus, the cosmological model that is accepted today by the majority of cosmologists, known as the "concordance model," would be impossible without inventing the intra-disciplinary field known as *astroparticle physics*. One should also note that the holy grail of theoretical physics—that is, constructing a consistent combination of quantum and relativity theories, an intra-disciplinary enterprise par excellence—is a *sine qua non* for a successful understanding of the very early history of the universe. On the next level of *interdisciplinary* transgression, cosmology exhibits an even richer case. Here, one important example is the full appropriation by cosmology of the techniques of computational statistics such as model fitting and Bayesian data analysis. Lastly, I define an *extra-disciplinary* transgression as the one between different domains of knowledge, which, in our case, occurs between cosmology, theology, and philosophy. This form of transgression presents itself not only in books by notable cosmologists such as Stephen Hawking, Lawrence Krauss, or Joseph Silk, but also in the debates held between these and other scientists on the one hand, and theologians and philosophers on the other (both in print and in publicly held meetings, mostly in university settings): Cosmologists debate fiercely with philosophers and theologians, claiming that their scientific jurisdiction licenses them to argue about topics such as the existence of God or the nature of nothingness.

I urge that these transgressions that cosmology engenders in the process of establishing its scientific legitimacy attest to the fact that a crucial aspect of the way scientific knowledge is usually characterized is being challenged today. This characterization, which found, as I argue, one of its best conceptualizations in Weber's classic *Wissenschaft als Beruf* (2011), is summarized in his famous "disenchantment" thesis, proposing that there exists a sharp boundary between the questions under the jurisdiction of science and the questions of "meaning" (*Sinn*), which science refuses to answer. As the current practice of science shatters this boundary, the assumptions pertinent to the social contexts within which it comes into existence will also be questionable. This, as I aim to show, is what we witness in the case of modern cosmology.

My paper is structured as follows: I begin with an exposition of Weber's account of scientific knowledge and scientific practice through an attentive reading of his *Wissenschaft als Beruf*. I mainly focus on Weber's defense of the

---

(Footnote 1 continued)
form of transgression possible. But to the extent that it exists in all interdisciplinary constructions, this criterion justifies my treatment of them as boundary transgressions.

disenchantment thesis and explore the justificatory structure of his argument. Next, I examine the tripartite structure of the boundary transgressions in current cosmology. I start with the *intra-disciplinary transgression* and investigate how one of the central pillars of modern cosmology, namely the *inflationary model* was born out of the intra-disciplinary formation of astroparticle physics. I then consider *interdisciplinary transgressions* and treat the statistical techniques cosmology employs in generating empirical evidence from data. In the last stage of the argument, I examine the *extra-disciplinary transgressions* that occur in cosmology in the context of the current philosophical and theological interventions of some of today's leading cosmologists. I argue that this form of transgression directly builds on the previous two and it has brought about a significant shift in the auto-representation of this science. I aim to demonstrate that it is the extra-disciplinary transgression, with a direct support from the previous ones, which carries cosmology beyond the Weberian dichotomy of science versus meaning and enables it to make claims that Weber believed to lie beyond the realm of scientific discourse. I suggest that without examining the recent combination of cosmology with M-theory, we will not be able to make sense of many pronouncements by authors such as Leonard Susskind or Stephen Hawking on topics such as the existence of God, the relationship between science and philosophy or the scientific status of the anthropic principle. In a final concluding section, after giving a review of the main argument, I consider some possible objections to my argument and ask to which extent the case in cosmology can represent a more general trend.

2   The Idea of a Weberian Science

The aim of this section is to give a layout of a view of scientific knowledge that Max Weber expounded in his lecture *Wissenschaft als Beruf* (originally published in 1917). I will first offer a detailed reading of Weber's argument and then explore how certain recent developments in cosmology compel us to re-think some of the premises of it.[2]

---

[2]I chose Weber's text as the background for my discussion since I believe that it provides a very succinct statement of a view on science that still prevails to this day. Although I cannot propose an argument here, to the extent that Popper's views on the tentative character of scientific knowledge constitute one of the major threads within mainstream philosophy of science, I contend that Weber's lecture can be seen as a source text for what has since followed. Specifically, as we see in the quote below, it is in the idea that *to be superseded*

In order to clarify the connection between Weber's lecture *Wissenschaft als Beruf* and what we observe in modern cosmology, I propose to focus on the way he uses the term "meaning" (*Sinn*) and how this term relates to his particular conceptualization of science. To begin with, let me point to the following passage where he introduces the crucial conceptualization in question, by referring to

> [...] the realm of science, where we all know that what we have achieved will be obsolete in ten, twenty, or fifty years. That is the fate, indeed, that is the very *meaning* of scientific work.[3] It is subject to and dedicated to this meaning in quite a specific sense, in contrast to every other element of culture of which the same might be said in general. Every scientific 'fulfillment' gives birth to new 'questions' and cries out to be surpassed and rendered obsolete [...] But we must repeat: to be superseded scientifically is not simply our fate but our goal. We cannot work without living in hope that others will advance beyond us. In principle, this progress is infinite (Weber 2004, p. 11).

Note that even though Weber seems to take it for granted that for a scientific achievement to be surpassed is equivalent for it to become "obsolete," this is only due to the fact that future science will be far more *advanced* than ours. In other words, future science will not simply be different, it will be *better*.[4] From this notion of scientific activity as "infinite progress," Weber immediately passes on to question whether such an enterprise which does not have an end can have any "point". He refers to this question as the "problem of the *meaning* of science" (or the meaning problem of science)[5] (Weber 2004, p. 11). He says:

> [...] it is far from self-evident that a thing that is subject to such a law can itself be *meaningful and rational*.[6] What is the point of engaging in something that neither

---

(Footnote 2 continued)
scientifically is not simply our fate but our goal that I see the seeds of what will later become the central thesis of Popper's philosophy of science. For the classic statement of Popper's philosophy, see Popper (2002). A recent influential reappraisal of Popperian ideas can be found in Mayo (1996).

[3]The original phrase is: "der *Sinn* der Arbeit der Wissenchaft" (Weber 2011, p. 15).

[4]This point becomes clear when we examine how Weber contrasts art with science. He writes that when, say, a technical development occurs in art, this does not mean that the art produced as a result of this development is "superior" or "higher" to early eras, whereas *in science, he thinks this is exactly the case*. See Weber (2004, p. 11).

[5]The sentence in German reads: "Und damit kommen wir zu dem *Sinnproblem* der Wissenschaft" (Weber 2011, p. 15).

[6]One should note that in this passage Weber's aim is to question the "point" of scientific activity and not to claim that it is "irrational". The German original reads: "*Denn das versteht sich ja doch nicht so von selbst, daß etwas, das einem solchen Gesetz unterstellt ist, Sinn und Verstand in sich selbst hat*" (Weber 2011, p. 15).

comes, nor can come to an end in reality? [...] What meaningful achievement can we hope for from activities that are always doomed to obsolescence? (Weber 2004, p. 12, emphasis added).

On the basis of these questions concerning the *meaning* (*Sinn*) of science, Weber introduces the general idea of "intellectual process of rationalization" (Weber 2004, p. 12) of modernity as follows: As we have just seen, the intrinsic nature of scientific activity, which is characterized as "infinite progress," makes the very *meaning* of this activity questionable: "For the man caught up in the chain of progress always has a further step in front of him [...]" (Weber 2004, p. 13). Still, Weber notes that this scientific rationalization process, which could be said to have begun centuries ago in the West, had an answer to this *question of meaning* before the advent of modernity. For both in its logical beginnings in Ancient Greek philosophy and in the empiricism that appeared in the Renaissance—i.e., in the idea of "rational experiment as a way of controlling experience reliably"—science had the meaning of being a "path to true *nature*" (Weber 2004, p. 15). This notion, under the influence of Protestantism, was later transformed into the idea of science as "the way to God" (Weber 2004, p. 16). Here, the idea was, by studying the findings of the natural sciences, one would "find clues to his [i.e., God's] intentions for the world" (Weber 2004, p. 16). For Weber, all this is in sharp contrast with the modern conception of science, according to which "science is alien to God" (Weber 2004, p. 16) and all those earlier conceptions of science such as "the path to true existence," "the path to true nature," or "the path to true God" are illusions that are being shattered (Weber 2004, p. 17).[7] For a modern person, Weber argues, it is childish to think that natural sciences can "teach us anything about the *meaning* of the world" (Weber 2004, p. 16).

But if science cannot answer the question of meaning—and Weber has no doubts on this, for he writes: "The fact that science cannot give us this answer is absolutely indisputable" (Weber 2004, p. 17)—then why is scientific activity *important*, in other words, worth pursuing? For him, *that* science is important can only be a *presupposition* which cannot be proven within scientific activity:

> It [i.e., the presupposition that science is important] can only be *interpreted* with reference to its ultimate meaning, which we must accept or reject in accordance with our own ultimate attitude towards life (Weber 2004, p. 18).

---

[7]Compare these observations of Weber with what we find in some of the modern cosmologists' popular books and lectures. For example, a well-known researcher and popularizer of cosmology, Paul Davies, who was a post-doctoral student of Fred Hoyle in Cambridge University, wrote: "[...] in my opinion science offers a surer path to God than religion" (Davies 1984, p. ix).

At this point, we obtain the argument structure of the *disenchantment thesis*. According to Weber, in each science there are presuppositions that cannot be accounted for by the science in question. This leads to the boundary that I referred to above as being between questions under the jurisdiction of science and the questions of meaning. The questions of meaning can only belong to the realm of presuppositions, with which science is incompetent to deal. Here, it is important to keep the structure of the argument in mind: Science is an activity which has the inherent tendency of progressing without limits and therefore cannot supply an end to be reached which would constitute its meaning. Therefore, the question of meaning *cannot* be posed as a scientific question.

I emphasize the link Weber sets up between the idea of "infinite progress" as the nature of scientific activity and the loss of meaning of the world for the modern human being. For I argue that the multidisciplinary nature of cosmology carried it beyond this understanding of science as a system of infinite progress whose goal is always to surpass itself. It is in this sense that the Weberian argument breaks down. Modern cosmology, combining the recent advanced observations of the matter-energy content of the universe, with the developments in string theory and its most ambitious version, namely the *M-theory*, which was originally proposed as a *theory of everything*, aims to outline a picture of the universe which is essentially right. To be sure, I do not argue that this aim is achieved or will be achieved in the near future. My argument concerns the practice of science and how cosmologists present their work in both scholarly publications and popular articles. I believe—as I document in the third section of this paper—there is a considerable body of evidence that modern theoretical cosmologists see their field as capable of achieving completion. In this regard, I argue that recent theoretical understanding of the universe gives us good reasons to think that the Weberian notion of the "infinite progress" of science is highly implausible.

Let me now go back to the question of "rationality" and "rationalization through science." I accept that the definition Weber gives for this term, viz., that "we are not ruled by mysterious, unpredictable forces, but that, on the contrary, we can in principle *control everything by means of calculation*" (Weber 2004, p. 13) is, in principle, still valid for contemporary cosmology. What I question is the assumption that this "rational science" will always stay within certain pre-given boundaries due to the idea of infinite progress that Weber brings to the fore. As I stressed above, once cosmology drops that assumption through the boundary transgressions that I describe below, "disenchantment due to rationalization" becomes questionable.

3   The Multidisciplinary Formations of Cosmology

The cosmological model that the majority of working scientists endorse today (otherwise known as the *concordance model*) is technically referred to as the Λ-CDM model. Here Λ refers to the cosmological constant that was first introduced by Albert Einstein to adjust his gravitational field equations in order to predict a stable universe (which later turned out to be an untenable assumption), and CDM stands for *cold dark matter*. The intriguing history of this Λ-term (see Earman 2001) led to its being appropriated, through the invention of the intra-disciplinary field of *astroparticle physics*, into standard cosmology. Though Einstein originally introduced the Λ-term into his field equations as a geometric entity, it was later re-interpreted within the astroparticle paradigm as a *vacuum energy density*. It is worth noting that the best estimate given for Λ, within the current astroparticle paradigm, is 120 orders of magnitude off than the observed value. One might argue that this shows the strength of cosmology's commitment to intra-disciplinarity: Although recognized as a problem, this discrepancy does not lead current cosmologists to repudiate the astroparticle interpretation.

Before, I referred to the transformation of cosmology, as it established itself as a legitimate science. I aim to argue that this transformation is intimately related to the boundary transgressions that occur in this context. I should like to point out that I do not aim to commit myself to a strong stance according to which one can determine the exact criteria as to what makes a discipline a "science," or that it can be established exactly when a discipline became established as one. The point of *transformation*, as I use this term, is not to pinpoint exact dates, but to draw attention to the fact that there are considerable differences in the way cosmology is practiced today as opposed to its earlier formations. Be it the initial attempts at model building in the works of Einstein, Willem de Sitter or Arthur Eddington at the turn of the twentieth century, or the work of George Gamow and his collaborators in the early 1950s, which paved the way for the discovery of the cosmic microwave background radiation (CMBR), we should recall that these efforts existed in an empirical environment which was significantly poorer compared to today's "scientific cosmology," and hence there were considerable difficulties in deciding between various cosmological models. In contrast, modern cosmology is a data-driven science based on a "standard model." A full account of this historical transformation is beyond the scope of this paper. Below, I will quote from a couple of influential textbooks to see how this transformation is represented by the discipline itself.

For example, Andrew Liddle's introduction to his undergraduate level textbook exemplifies the confidence that distinguishes contemporary cosmology:

> The development of cosmology will no doubt be seen as one of the scientific triumphs of the twentieth century. At its beginning, cosmology hardly existed as a scientific discipline. By its end, the Hot Big Bang cosmology *stood secure as the accepted description of the Universe as a whole*. Telescopes such as the Hubble Space Telescope are capable of seeing light from galaxies so distant that the light has been traveling towards us for most of the lifetime of the Universe. The cosmic microwave background [...] is routinely detected and its properties examined. That our Universe is presently expanding is *established without doubt* (Liddle 2003, p. xi, emphasis added).

In his widely used textbook, Scott Dodelson also emphasizes the "quantitative turn":

> There are two aspects of cosmology today that make it more alluring than ever. First, there is an enormous amount of data. To give just one example of how rapidly our knowledge of the structure of the universe is advancing, consider galaxy surveys which map the sky. In 1985, the state-of-the-art survey [...] consisted of the positions of 1100 galaxies. Today, the Sloan Digital Sky Survey and the Two Degree Field between them have recorded the 3D positions of half a million galaxies.

> *The other aspect of modern cosmology which distinguishes it from previous efforts to understand the universe is that we have developed a consistent theoretical framework which agrees quantitatively with the data.* These two features are the secret of the excitement in modern cosmology: we have a theory which makes predictions, and these predictions can be tested by observations (Dodelson 2003, p. x, emphasis added).

According to Dodelson, the "new" cosmology not only makes precise humanity's traditional questions concerning the universe, it can also test them:

> The realization that the universe is expanding and was once much hotter and denser allows us to modernize the deep age-old questions 'Why are we here?' and 'How did we get here?' The updated versions are now 'How did the elements form?', 'Why is the universe so smooth?', and 'How did galaxies form from this smooth origin?'. Remarkably, these questions and many like them have quantitative answers, answers that can be found only by combining our knowledge of fundamental physics with our understanding of the conditions in the early universe. Even more remarkable, these answers can be tested against astronomical observations (Dodelson 2003, p. 1).

These remarks are typical. To move to another medium from textbooks, one reads on the web page of a recently founded interdisciplinary research center, the Institute for Strings, Cosmology, and Astroparticle Physics (ISCAP), at Columbia University:

This is the golden age of cosmology. Once a data-starved science, cosmology has burgeoned as ground and space-based astronomical observations supply a wealth of unprecedentedly precise cosmological measurements. Questions that were recently the stuff of speculation can now be analyzed in the context of rigorous, predictive theoretical frameworks whose viability is determined by observational data. Finally, cosmological theory is being confronted by cosmological fact. The most surprising and exciting feature of cosmology's entrance into the realm of data-driven science is its deep reliance on theoretical developments in elementary particle physics. At the energy scales characteristic of the universe's earliest moments, one can no longer approximate matter and energy using an ideal gas formulation; instead, one must use quantum field theory, and at the highest of energies, one must invoke a theory of quantum gravity, such as string theory. Cosmology is thus the pre-eminent arena in which our theories of the ultra-small will flex their muscles as we trace their role in the evolution of the universe (ISCAP 2015).

Note that the last quote combines the transformation of cosmology with the need of intra-disciplinary expansion. Many other examples can be found in the literature to the same effect: Cosmology presents itself as a mature, quantitative science which has a right to exercise a technical sovereignty over ancient questions such as "Why are we here?" In the next section, I demonstrate how this scientific status is achieved through various forms of multidisciplinary transgressions.

3.1 Intra-Disciplinary Transgression: The Inflationary Universe

I previously mentioned that the inflationary model is one of the integral components of the ΛCDM model. Here I aim to explicate how inflation came into existence as a result of an intra-disciplinary transgression, and to this effect, I will look at its historical roots in the work of Alan Guth.

Inflation was introduced by Guth, then a particle physicist, in 1981. In his seminal paper (Guth 1981), he argued that inflation solves three major problems faced by the standard Big Bang model, namely the *horizon, flatness*, and the *monopole* problems.[8] Here, without going into technical details, I want to explicate the role played by intra-disciplinary considerations in his proposal. Let me first explain the three problems in a qualitative way.

---

[8]Detailed discussions of these problems can be found in almost any textbook on contemporary cosmology. For example, see Dodelson (2003). A very readable non-technical account is given by Guth (1998).

Horizon problem: When we observe the CMBR from different parts of the sky, we see that their temperatures are almost exactly the same. This is puzzling for when these photons that we observe now were radiated, they were too far away from each other to be in causal contact. Hence, their temperature could not have equalized through an interaction. The classical Big Bang Model (BBM) simply posited that this uniformity must come from the initial conditions of the universe. However, this way of explaining a uniformity through "fine-tuning"[9] is not satisfactory.

Flatness problem: Our observations confirm that the universe is very nearly flat, i.e., at the very large scale, Euclidean geometry correctly describes space. However, it can be shown that a flat universe is not stable; a small perturbation could have made the universe deviate from flatness and become curved. As we do not observe this, BBM had to assume that the universe began extremely fine-tuned so that it could stay flat in its entire history.

Monopole problem: A magnetic monopole is a hypothetical particle which would act as a magnetic charge carrier, similar to an electron as the electric charge carrier. They have never been observed experimentally, even though they have been predicted to exist in *grand unified theories*. As these theories are believed to describe the early universe also, we expect to observe these particles.

Guth's original research was not on modeling the early universe but on particle physics, and in particular, certain problems that arose in the context of *grand unified theories* or GUTs. The aim of a GUT is to give a unified theory of three fundamental interactions in the universe, excluding gravity. In other words, it aims to create a single quantum theory of electromagnetic, strong and weak forces. In terms of fields of study, this means combining quantum electrodynamics with quantum chromodynamics and the quantum theory of weak interactions.[10] What

---

[9]That is to say, the BBM had to assume that the universe started in a way that the temperature everywhere was exactly the same as if it was pre-arranged.

[10]The program of unification in fundamental physics has a long history going back at least to Maxwell's work on electricity and magnetism. In the context of particle physics, the most fruitful result achieved up until today is the unification of the weak and electromagnetic interactions by Abdus Salam, Sheldon Glashow, and Steven Weinberg, who were awarded the Nobel Prize in Physics in 1979 for their work. At present, no such success seems to be on the horizon for the GUT project. I recognize that the question whether unification can be analyzed as a boundary transgression is an important one but it is beyond the limits of this paper. Still, it seems reasonable to think that it should be understood as an intra-disciplinary formation, for in unification one combines two or more subdisciplines with their own separate set of entities, questions, and methods. Even though in the context of particle physics there are considerable similarities between the subdisciplines that one aims to unify,

Guth noticed was that a possible physical scenario which is worked out within the context of GUT models can be invoked to deal with certain problems within the "old" Big Bang theory. As I mentioned above, the connection between the BBM and GUTs is through the monopole problem: Even though the GUTs predict that plenty of magnetic monopoles should exist, we do not observe them. The solution that Guth came up with, viz., the inflationary mechanism, to explain the nonexistence of monopoles also solves the horizon and flatness problems without resorting to fine-tuning. I should note that Guth presented his argument putting the horizon and flatness problems at the forefront and only occasionally mentioned that inflation can also provide a mechanism to solve the monopole problem. It is instructive to read from the abstract of his paper:

> The standard model of hot big-bang cosmology requires initial conditions which are problematic in two ways: (1) The early universe is assumed to be highly homogeneous, in spite of the fact that separated regions were causally disconnected (horizon problem); and (2) the initial value of the Hubble constant must be fine tuned to extraordinary accuracy to produce a universe as flat (i.e., near critical mass density) as the one we see today (flatness problem). These problems would disappear if, in its early history, the universe supercooled to temperatures 28 or more orders of magnitude below the critical temperature for some phase transition. A huge expansion factor would then result from a period of exponential growth, and the entropy of the universe would be multiplied by a huge factor when the latent heat is released. *Such a scenario is completely natural in the context of grand unified models of elementary particle interactions. In such models, the supercooling is also relevant to the problem of monopole suppression* (Guth 1981, p. 347, emphasis added).

In very basic terms, what inflation says is that the universe went through a phase transition in its early history which led to a huge expansion. This solves the horizon problem for now all the radiation that we receive from CMBR can be considered as originating from a single causal "patch," which became the entire observable universe around us after the expansion ended.[11] The flatness problem is also solved for what this expansion does is to take a very small curved portion of the universe and smooth it out to become flat. In both cases, an explanatory mechanism replaces a brute assumption of finely tuned initial conditions. Finally, as the expansion

---

(Footnote 10 continued)
the differences are far from trivial, which explains why, as I mentioned, a consistent and empirically adequate GUT does not yet exist.

[11]In other words, as inflation gives an account of how a small enough space that was causally connected in the early universe could become the entire observable universe, the homogeneity of temperature is not puzzling. There was enough time for radiation to equalize the temperature.

makes a tiny volume into a huge one, we do not expect many monopoles to exist in the final volume, given that it came from a tiny one which probably only included a very small amount of monopoles.

The intra-disciplinary structure is unmistakable: Inflation is motivated within the "context of grand unified models of elementary particle interactions." These GUT models, which are intra-disciplinary on their own, are in turn applied within the context of cosmology. Guth is aware of the transgression he is engendering. For after introducing the horizon problem as follows: "The initial universe is assumed to be homogeneous, yet it consists of [...] separate regions which are causally disconnected (i.e., these regions have not yet had time to communicate with each other via light signals)" (Guth 1981, p. 355), he adds in a footnote:

> In order to calculate the horizon distance, one must of course follow the light trajectories back to t = 0. This violates my contention that the equations are to be trusted only for $T \leq T_0$. Thus the horizon problem could be obviated if the *full quantum gravitational theory had a radically different behavior from the naive extrapolation [...] However, it is the point of this paper to show that the horizon problem can also be obviated by mechanisms which are more within our grasp, occurring at temperatures below T_0* (Guth 1981, p. 355, emphasis added).

For the astroparticle physics project to get off the ground, certain contentious assumptions have to be put in place. Thus one has to assume that the extrapolation to the very early universe of the standard model is secure. Without this assumption, the horizon problem could not even be defined within the context of the paper and the argument would lose a major motivational support. In order to bring the physics of the early universe within the jurisdiction of the GUT program, an intra-disciplinary transgression occurs which pictures the early universe as an astroparticle "laboratory."

3.2 Interdisciplinary Transgression: Discovery as Model Selection

The importance of the level of interdisciplinarity in the context of cosmology and statistics stems from the fact that without this merging, it would be impossible to transform data to evidence. To be sure, statistical methods and computational techniques are used in many diverse fields today. What is important for my purposes is not *that* these methods are used, but *how* cosmology implements them. The case of Bayesian methods is especially worth discussing because they can only be applied on the basis of some contentious assumptions. Here, I aim to focus on one particular application of these techniques: the discovery of the acceleration of the universe.

I will examine, albeit briefly, the work of two research teams that successfully devised new experimental techniques and methods of analysis to overcome the obstacles that cosmological distance measurements presented since the time of Hubble. I aim to explicate how the empirical dynamics that eventually led to both teams'—namely, the High-Z Supernova Search Team (hereafter, HZS) and the Supernova Cosmology Project (hereafter, SCP)—receiving credit for the discovery of the acceleration of the universe depended on certain statistical considerations. This discovery, which was later interpreted in terms of a still not completely understood notion of "dark energy," forms an integral part of the modern *concordance model* in cosmology.

The central question that I want to pose can be formulated in the following way: Given that it is generally accepted that the expansion of the universe is accelerating, how is this conclusion justified empirically? In order to deal with this question, I will focus on the evidence papers, following Staley (2011), that the groups produced.[12] I suggest that the best way to address this question is through approaching the experimental program that governs the measurements of both teams within the context of *model selection* problem in *statistical inference theory*. In this setting, it appears that the epistemology of experiment, to use a term introduced by Allan Franklin, defined as the study of "how we come to believe rationally in an experimental result" (Franklin 1989, p. 165), that led the researchers to conclude that the expansion of the universe is accelerating depends crucially on an epistemic structure with two main components, both of which are based on the notion of the *elimination of systematic errors*:

1. Error elimination in observations through identification of bias and contentious assumptions.
2. Error elimination in data analysis through improved statistical inference schemes.

Furthermore, tasks to meet the requirements of each of these epistemological components were carried out with an underlying statistical tool that is known as *robustness analysis*. Consequently, the following thesis emerges as the key to the epistemology of the experiments conducted by HZS and SCP: The experimental

---

[12]These being Perlmutter et al. (1999) for the SCP collaboration and Riess et al. (1998) for the HZS.

work that resulted in the discovery of the accelerating universe is best characterized as an experimental model selection effort based on constraining cosmological parameters statistically, which is achieved by the strategy of error elimination that employs robustness analyses in various forms.

In line with the fact that the main justificatory argument for the experimental result is based on *statistical inference*, both teams aim at finding out which cosmological models are ruled out by the data on statistical grounds and which models are consistent with it. For both teams, the empirical validity of their results comes from their ability to contain through statistical means the adverse effects of the systematic uncertainties that are present in their data due to various biases and other astrophysical problems. That is to say, even though neither of the teams can eliminate systematic errors fully, they are still able to argue that they can discriminate among competing models, for the favored model is *robustly* supported. Here *robustness* is generally understood as: the value of a parameter remaining consistent with the result when the fitting method or the data points included in the analysis are changed.

Thus, HZS explains their aims as:

> Measurement of the elusive cosmic parameters Ω_M and Ω_Λ through the redshift-distance relation depends on comparing the apparent magnitudes of low-redshift SNe Ia with those of their high-redshift cousins [...].

> The High-Z Supernova Search Team has embarked on a program to measure supernovae at high redshift and to develop the comprehensive understanding of their properties required for their reliable use in cosmological work (Riess et al. 1998, p. 1010).

The data they collect consists of spectral and photometric observations of 34 supernovae. Once this data is gathered, the question becomes: What does it imply about the fundamental cosmological parameters Ω_M and Ω_Λ, which, respectively, stand for the density of matter in the universe and the cosmological constant or vacuum energy density, in the language of astroparticle physics. The problem thereby turns into: Given the data, what is the probability of, say, Ω_M to have a particular value? At this stage, HZS resorts to Bayes' theorem, which is a mathematical formula that computes the desired probability in terms of several prior probabilities. As in most cases there is no independent way of ascertaining those prior probabilities, scientists must use "informed guesses." HZS puts the issue as follows:

Because the values of A_v[13] are not known a priori, we use an initial guess derived from the color excess measured from the uniform color range of SNe Ia (Riess et al. 1998, p. 1036).

This apparently subjective method involved in Bayesian analysis makes it suspect in the eyes of many philosophers of science and some statisticians as well. However, for HZS, *robustness*, understood as the agreement of the values of the parameters obtained through different methods, is a way out of this dilemma. Thus they employ the two methods of light-curve fitting, viz., the template-fitting method and the multicolor light curve shape method. The employment of these methods is crucial for HZS to claim *robust* statistics for their results.

The case for the SCP collaboration is quite similar. Interestingly, they employ both Bayesian and frequentist methods and provide 12 different fits, which are all strongly inconsistent with a non-accelerating universe. These variations in fittings constitute the basis for the *robustness claim* of the SCP team. Their result is expressed in strictly statistical terms, carefully distinguishing statistical from systematic errors:

> The measurement yields a joint probability distribution of the cosmological parameters that is approximated by the relation $0.8 \ \Omega_M - 0.6 \ \Omega_\Lambda \sim -0.2 \pm 0.1$ in the region of interest ($\Omega_M \leq 1.5$). For a flat ($\Omega_M + \Omega_\Lambda = 1$) cosmology we find $\Omega_M^{flat} = 0.28_{-0.08}^{+0.09}(1\sigma \ statistical)_{-0.04}^{+0.05}$ (identified systematics). The data are strongly inconsistent -0.04 with a $\Lambda = 0$ flat cosmology, the simplest inflationary universe model. An open, $\Lambda = 0$ cosmology also does not fit the data well: the data indicate that the cosmological constant is nonzero and positive, with a confidence of P ($\Lambda > 0$) = 99 %, including the identified systematic uncertainties (Perlmutter et al. 1999, p. 565).

They also add: "The size of our sample allows us to perform a variety of statistical tests to check for possible systematic errors and biases" (Perlmutter et al. 1999, p. 565). In fact, most of the paper is devoted to statistical analyses of various systematic uncertainties and cross checks. Comparing this with Edwin Hubble's early work (see Hubble 1929) on the expansion of the universe, as a representative of the "old" cosmology, one cannot help but notice the dominance of statistical methods in today's experimentalists' work. In establishing his result, Hubble did not resort to any statistical analysis at all.

Through its full appropriation of statistical methods, cosmology transgressed its boundary in observational astronomy and secured an epistemic domain on this

---

[13]Here, A_v stands for "extinction," which is the reduction in the intensity of light as it passes through the interstellar medium.

basis. It is this ability to establish empirical results that leads to model selection that Dodelson and Liddle have in mind when they refer to "precision cosmology" (see Liddle 2003, p. 117 and Dodelson 2003, p. 385).

3.3 Extra-Disciplinary Transgression: Cosmology Contra Philosophy and Theology

The third form of transgression that I will examine builds on the previous two, in the sense that it puts to good use the success of them.[14] The intra-disciplinary form of transgression enabled cosmology to construct a "standard" model of the universe, and the interdisciplinary form facilitated the confirmation of the model. On the basis of these developments, several prominent cosmologists carry out an extra-disciplinary transgression and put forward claims with theological and philosophical resonances. In other words, these theological or philosophical pronouncements are directly connected to this recent transformation in cosmology. I will examine the works of three prominent cosmologists to argue my point. We will see that in each of the cases, the authors in question build on their scientific work to reach conclusions which would traditionally be classified as belonging to extrascientific domains.

Before passing on to the discussion of the works of cosmologists, there are three concepts that I need to explicate briefly:

Fine-tuning problem: The notion of fine tuning can, in its most general form, be defined as the situation in which the stability of a structure depends on a certain parameter having a precise value,[15] without there being any natural explanation why this should be so. This structure could be a galaxy, the cosmic background microwave radiation, or life itself. Two well-known fine-tuning problems in cosmology are the flatness of the universe and the value of the cosmological constant (which, in turn, determine various structures in the universe).

---

[14]I should point out that each form of transgression is equally important for the main argument of this paper. Still, it will be noticed that I allocate more space to the third form of transgression compared to the previous two. This is both because the demonstration of the final transgression requires more textual resources than the others and also it is in the third form of transgression that we fully observe the effect which constitutes the subject matter of this paper, namely, the transformation in cosmology which brought certain questions of "meaning" within the jurisdiction of this science.

[15]As if the value of the parameter is "tuned" by a "purposeful designer."

Anthropic principle: The anthropic principle is best viewed as an attempt to solve the fine-tuning problem. In effect, it says that if the values of these universal parameters[16] were different, universe would not give rise to human beings to observe these very parameters.

Multiverse: The idea of multiverse arises through several routes, the most prominent ones being the string theory (which aims to combine quantum and relativity theories) and inflationary cosmology. According to the multiverse picture, our universe is only one among many others and the otherwise puzzling values of the cosmological parameters are explained by a statistical argument. In other words, the precise values of cosmological parameters are not a result of a fine-tuning process, for all possible values actually do obtain and they are distributed to each universe. We just happen to find ourselves in this particular universe, for it is this one (and here the anthropic selection principle is invoked) that makes our existence possible.

In what follows, I will argue that the debates concerning the origin of the universe out of nothing or the existence of human beings being used as a *natural*[17] explanation of our universe indicate a shift in the discourse of science and scientific knowledge. To repeat, my claim is that the seemingly "metaphysical" ideas propounded by Krauss, Susskind, Hawking, and others are actually traceable to physics articles published in peer-reviewed physics journals. It is on the basis of this reading of the cosmological problems as they are discussed in scientific journals and represented in popular magazines and books that I will make my concluding point concerning why we should re-think the disenchantment thesis.

3.3.1 Krauss on Nothingness and the End of Cosmology

Lawrence Krauss is currently Foundation Professor of the School of Earth and Space Exploration, and director of the Origins Project at the Arizona State University. In addition to his publications in scientific journals, he is also the author of several popular books on cosmology, the latest being *A Universe from Nothing (2012)*. In this section, I will analyze two particular arguments in this book.

The first argument is summarized in the title: Current cosmology can explain the very existence of the universe without the need of invoking a creator God. That

---

[16]Such as the charge of the electron or the value of the cosmological constant.

[17]That is to say, the anthropic principle is required to explain the universe that we live in naturally, without invoking the existence of a "creator."

is to say, quantum processes combined with inflation and dark energy can explain how a universe can come into existence from "nothing."[18] As he puts it:

> [...] if the quantum properties of matter and radiation end up endowing even an infinitesimally small region of empty space with energy at very early times, this region can grow to be arbitrarily large and arbitrarily flat. When the inflation is over, one can end up with a universe full of stuff (matter and radiation), and the total Newtonian gravitational energy of that stuff will be as close as one can ever imagine to zero. A universe from Nothing, indeed (Krauss 2012, p. 104).

The crucial point is: According to Krauss, staying entirely within the limits of modern cosmology, one can construct a "creation out of nothing" story in a scientifically legitimate way. Here, I will not go into many critical reviews the book, and this argument in particular, received. I rather wish to mention one exchange between Columbia University philosopher David Albert and Krauss on latter's use of "nothing," to illustrate my point. Krauss complains:

> Before going further, I want to devote a few words to the notion of 'nothing'—a topic that I will return to at some length later. For I have learned that, when discussing this question in public forums, nothing upsets the philosophers and theologians who disagree with me more than the notion that I, as a scientist, do not truly understand 'nothing.' (I am tempted to retort here that theologians are experts at nothing.)
>
> 'Nothing,' they insist, is not any of the things I discuss. Nothing is 'nonbeing,' in some vague and ill-defined sense [...] some philosophers and many theologians define and redefine 'nothing' as not being any of the versions of nothing that scientists currently describe. But therein, in my opinion, *lies the intellectual bankruptcy of much of theology and some of modern philosophy* (Krauss 2012, pp. xiii–xiv, emphasis added).

Yet, he wants to have none of this. He speaks with full scientific authority:

> One thing is certain, however. The metaphysical 'rule,' which is held as an ironclad conviction by those with whom I have debated the issue of creation, namely that 'out of nothing nothing comes,' has no foundation in science. Arguing that it is self-evident, unwavering, and unassailable is like arguing, as Darwin falsely did, when he made the suggestion that the origin of life was beyond the domain of science by building an analogy with the incorrect claim that matter cannot be created or destroyed. All it represents is an unwillingness to recognize the simple fact that *nature may be cleverer than philosophers or theologians* (Krauss 2012, p. 174, emphasis added).

---

[18]The conceptual and technical details of this argument are beyond the scope of this paper. As my aim is not to assess Krauss's argument but simply to document how it represents cosmology, I will be content with stating it in most general terms.

In a review published in *The New York Times*, Albert took direct aim at these remarks. In his usual terse style, he wrote, concerning Krauss's story of "creation out of nothing":

> But that's just not right [...] the fact that particles can pop in and out of existence, over time, as those fields rearrange themselves, is not a whit more mysterious than the fact that fists can pop in and out of existence, over time, as my fingers rearrange themselves. And none of these poppings—if you look at them aright—amount to anything even remotely in the neighborhood of a creation from nothing (Albert 2012).

He thus concluded:

> But all there is to say about this, as far as I can see, is that Krauss is dead wrong and his religious and philosophical critics are absolutely right (Albert 2012).

The second argument that I will focus on has its origin in Krauss's own work done with his collaborator, Robert Scherrer, a cosmologist at Vanderbilt University. In their joint paper (Krauss and Scherrer 2007), they presented an argument which implies that, due to the expansion of the universe, future observers will fail to receive any light from galaxies in the universe. What this means is that

> [...] as we extrapolate the current ΛCDM universe forward in time, all evidence of the Hubble expansion will disappear, so that observers in our 'island universe' will be fundamentally incapable of determining the true nature of the universe, including the existence of the highly dominant vacuum energy, the existence of the CMB, and the primordial origin of light elements. With these pillars of the modern Big Bang gone, this epoch will mark the end of cosmology and the return of a static universe (Krauss and Scherrer 2007, p. 1545).

Thus cosmology shows that we are extremely lucky to live in the current epoch of the universe. The ambiguity in his usage of the word "cosmology" (is it the fact that no cosmological event will occur due to the static existence of the universe or that no science of cosmology will be possible?) is quite harmless in the sense that both meanings would make the sentence true. In any case, the following quote shows that Krauss seems to think that the very science of cosmology will cease to exist:

> Observers when the universe was an order of magnitude younger would not have been able to discern any effects of dark energy on the expansion, and observers when the universe is more than an order of magnitude older will be hard pressed to know that they live in an expanding universe at all, or that the expansion is dominated by dark energy. By the time the longest lived main sequence stars are nearing the end of their lives, for all intents and purposes, the universe will appear static, and all evidence that now forms the basis of our current understanding of cosmology will have disappeared (Krauss and Scherrer 2007, pp. 1549–1550).

Krauss finds this fact astonishing. In *A Universe from Nothing*, after quoting from the paper a sentence he also repeats in his popular talks, namely, "We live at a very special time [...] the only time when we can observationally verify that we live at a very special time" (Krauss and Scherrer 2007, p. 1549), he remarks that:

> We were being somewhat facetious, but it is sobering to suggest that one can use the best observational tools and theoretical tools at one's disposal and nevertheless come up with a completely false picture of the large scale universe (Krauss 2012, p. 118).

Above, I pointed out that Weber understood the development of science as an infinite progress, i.e., science continuously gets better. However on the basis of Krauss's arguments, it appears that this idea of infinite progress depends on a certain understanding of cosmology, which is not possible to maintain any longer today. Current cosmology implies that the notion of infinite progress through the continuous application of scientific method is itself rendered obsolete, in the sense that it will be the future science that will get things wrong and not today's.

3.3.2 Hawking on M-Theory, God and Philosophy

I want to mention Stephen Hawking's case briefly, for his approach is considerably similar to Krauss's. In a series of papers (see Hawking 1983, 1984; Hawking and Hartle 1983), Hawking sought a way of applying quantum mechanics to the entire universe, a field known as quantum cosmology. On the basis of his work in quantum cosmology and M-theory, he argued in his popular books and lectures that physics can explain the origin of the universe naturalistically. For example, in his recent book, *The Grand Design*, after remarking that "M-theory is the only model that has all the properties we think the final theory ought to have" (Hawking 2011, p. 8), Hawking continues:

> M-theory predicts that a great many universes were created out of nothing. Their creation does not require the intervention of some supernatural being or god. Rather, these multiple universes arise naturally from physical law. They are a prediction of science (Hawking 2011, pp. 8–9).

Similar to Krauss, Hawking also found the occasion to dismiss philosophy in this work. Questions such as "How can we understand the world in which we find ourselves? How does the universe behave? What is the nature of reality?" he wrote, traditionally belonged to

> [...] philosophy, but philosophy is dead. Philosophy has not kept up with modern developments in science, particularly physics. Scientists have become the bearers of the torch of discovery in our quest for knowledge (Hawking 2011, p. 5).

I will remark further on Hawking's argumentation in *The Grand Design* in my conclusion. At this stage, I simply want to observe that Hawking thinks that his transgression into theology and philosophy with a single stroke is fully justified by M-theory-based cosmology.

3.3.3 Susskind on Multiverse and the Anthropic Principle

Leonard Susskind is presently the Felix Bloch Professor of Theoretical Physics at Stanford University and Director of the Stanford Institute for Theoretical Physics. His main research is on string theory and black hole physics. He is also one of the main proponents of the *anthropic principle* in the current scene of theoretical cosmology. This principle is subject to continuous debate, both within and outside cosmology and hence the extra-disciplinary transgression that takes place in the context of anthropic principle is a complicated one, being disputed both by philosophers and theologians (see, e.g., Bostrom 2007; Craig 1988). As George Ellis reports in his editorial note to Carter's classic *Large Number Coincidences and the Anthropic Principle in Cosmology* (Carter 1974):

> The anthropic principle is one of the most controversial proposals in cosmology. It relates to why the universe is of such a nature as to allow the existence of life. This inevitably engages with the foundations of cosmology, and has philosophical as well as technical aspects. The literature on the topic is vast—the Carter paper reprinted here [...] has 226 listed citations, and the Barrow and Tipler book [...] has 1740 (Ellis 2011, p. 3213).[19]

Again, without going into the technical intricacies of the anthropic argument, I want to argue that Susskind's employment of the principle is in line with the general form of boundary transgressions that I have been outlining in this paper. As he explains in a heavily cited paper (Susskind 2007),[20] Susskind arrives at the anthropic principle via string theory. String theory, which originated in particle physics as an explanation for quark confinement, was later re-interpreted as a theory of quantum gravity, i.e., a theory which combines general relativity and quantum mechanics. As is well known, even though the theory is mathematically very sophisticated, many physicists think that it failed to fulfil its premise as a scientific theory for it could not produce any predictions that could be tested experimentally. One striking feature of the mathematical complexity of the theory is that it gives rise to an enormous number of possible solutions, the so-called *landscape*. Given so many possibilities, the answer why this particular universe

---

[19]The book by Barrow and Tipler that Ellis mentions is Barrow and Tipler (1988).

[20]This paper received 561 citations at the time of this writing.

rather than any other is answered by invoking the anthropic constraint: because this is the type of universe that gave raise to us. As Susskind explains:

> [...] in an anthropic theory simplicity and elegance are not considerations. The only criteria for choosing a vacuum [i.e., a solution] is utility, i.e. does it have the necessary elements such as galaxy formation and complex chemistry that are needed for life. That together with a cosmology that guarantees a high probability that at least one large patch of space will form with that vacuum structure is all we need (Susskind 2007, p. 252).

Susskind does not seem to question string theory. Instead, he accepts the paradigm and argues that the anthropic principle is an inevitable piece of the puzzle:

> With nothing preferring one vacuum over another, the anthropic principle comes to the fore whether or not we like the idea. String theory provides a framework in which this can be studied in a rigorous way.

> Progress can certainly be made in exploring the landscape [...] We can argue the philosophical merits of the anthropic principle but we can't argue with quantitative information about the number of vacua with each particular property such as the cosmological constant, Higgs mass or fine structure constant (Susskind 2007, p. 262).

Most critiques of the anthropic principle used the "falsifiability" criterion to attack it. Susskind answers them as follows:

> Good scientific methodology is not an abstract set of rules dictated by philosophers. It is conditioned by, and determined by, the science itself and the scientists who create the science. What may have constituted scientific proof for a particle physicist of the 1960's—namely the detection of an isolated particle—is inappropriate for a modern quark physicist who can never hope to remove and isolate a quark. Let's not put the cart before the horse. Science is the horse that pulls the cart of philosophy (Susskind 2006, p. 194).

And a couple of lines later:

> As for rigid philosophical rules, it would be the height of stupidity to dismiss a possibility just because it breaks some philosopher's dictum about falsifiability. What if it happens to be the right answer? (Susskind 2006, p. 196).

I tend to read these lines as exemplifying boundary transgression, in the sense I describe in the body of this paper, for the following reason. The problem that Susskind attempts to solve is a strictly scientific one: providing a physical reason why instead of all possible mathematical solutions in the *string landscape, we observe one particular universe.* The solution he offers via the anthropic principle raises critical methodological worries from the philosophy of science perspective. At this point, instead of facing the philosophical criticism in its own domain,

Susskind prefers to reject the question through transgressing the boundary between science and philosophy: If the anthropic principle is successful, there is no need to answer the philosophical question for "science is the horse that pulls the cart of philosophy."

4 Conclusion

The structure of the argument that I attempted here has four main moments:

1. Cosmology went through what I named a *transformation process.*
2. This process was built on various multidisciplinary interactions.
3. These multidisciplinary interactions realized various forms of boundary transgressions.
4. Finally, these boundary transgressions enabled cosmology to reach conclusions which carried it beyond the Weberian picture of science, the effects of which can be seen in the auto-representations of cosmology in the writings of prominent authors.

Throughout the paper, I have been at pain to stress that various philosophical and theological pronouncements of cosmologists can be traced back to the multidisciplinary developments that characterize modern cosmology. Without the application of Bayesian statistical arguments, cosmologists would not be able to empirically ascertain the existence of dark energy. Also, without the confident invention and application of astroparticle physics to the early universe, they would not be able to argue that the universe arose "out of nothing."

One may still ask whether I put too much weight on popular science books written by a handful of senior scientists to make my case. To which extent do my claims represent the majority view of cosmologists? To be sure, it is impossible to give a fully satisfactory answer to this question without applying quantitative social scientific methods to intellectual history. But to the extent that these arguments provided by Susskind, Krauss and Hawking can be *rationally* explained, as I tried to do, we might claim that their representative force goes beyond their statistical magnitude.

In what sense does cosmology signal the "end" of Weberian science? Above, I promised to remark further on Hawking. One theme that he brings up in his discussion is the notion of a *final theory.* Provided that physicists are confident that there are only four types of forces in nature, it is reasonable for them to think that a theory which can explain them in a unified way would be the theory of everything.

In this way, physics can have an "end." But is this what Weber had in mind? If the possibility of constructing the final theory (which would also complete cosmology) that would explain the world staying fully within naturalistic premises that Krauss and Hawking dream of is open, one may expect that this would have serious theological and philosophical repercussions. Weber argued that science cannot deal with the question of meaning, assuming its infinite progress. However, we have seen that the boundary transgressions of cosmology carried several notable practitioners of this science to deal with issues that Weber assumed to lie forever outside the scope of science, and to do this on the very basis of cosmology. One other concern in this regard is Krauss's observation that a knowledge horizon will exist for the future observers of the universe due to the accelerated expansion. Once the Weberian idea of an infinitely progressing science is put into question, Dodelson's translation of a question such as "Why are we here?" into "How did the elements form?" might become more than a rhetorical device used to open a textbook.

Weber delivered his lecture in 1917, when we still "lived" in a static universe. It was not until Hubble's (1929) discovery of the galaxy red-shift velocity relation that the notion of a "dynamic" universe became a real possibility. Considered from a historical perspective, an evolving universe is a major paradigm change. Perhaps it is not that surprising that this eventually brings about repercussions on scientific knowledge itself to re-enchant our world as we cognize the universe in unexpected ways.

References

Albert D (2012) On the origin of everything: 'A Universe From Nothing,' by Lawrence M. Krauss. http://www.nytimes.com/2012/03/25/books/review/a-universe-from-nothing-by-lawrence-m-krauss.html. Accessed 18 June 2015

Barrow J, Tipler F (1988) The anthropic cosmological principle. Oxford University Press, Oxford

Bostrom N (2007) Anthropic bias: observation selection effects in science and philosophy. Routledge, London

Carter B (1974) Large number coincidences and the anthropic principle in cosmology. In: Longair MS (ed) Confrontation of cosmological theories with observational data. Proceedings of the symposium, Krakow. D. Reidel, Dordrecht, pp 291–298

Craig WL (1988) Barrow and Tipler on the anthropic principle vs. divine design. Br J Philos Sci 39:389–395

Davies P (1984) God and the new physics. Simon & Schuster, New York

Dodelson S (2003) Modern cosmology. Academic Press, Boston

Earman J (2001) Lambda: the constant that refuses to die. Arch Hist Exact Sci 55:189–220

Ellis G (2011) Editorial note to: Brandon Carter, large number coincidences and the anthropic principle. Gen Relat Gravit 43:3213–3223

Franklin A (1989) The neglect of experiment. Cambridge University Press, Cambridge

Guth AH (1981) Inflationary universe: a possible solution to the horizon and flatness problems. Phys Rev D 23:347–356

Guth AH (1998) The inflationary universe. Basic Books, Boston

Hawking SW (1983) Quantum cosmology. In: DeWitt B, Stora R (eds) Relativity, groups and topology. Les Houches, Amsterdam

Hawking SW (1984) The quantum state of the universe. Nucl Phys B 239:257–285

Hawking SW (2011) The grand design. Bantam, London

Hawking SW, Hartle JB (1983) The wave function of the universe. Phys Rev D 28:2960–2975

Hubble E (1929) A relation between distance and radial velocity among extra-galactic nebulae. Proc Natl Acad Sci USA 15(3):168–173

ISCAP—Institute for Strings, Cosmology, and Astroparticle Physics (2015) About ISCAP. http://www.iscap.columbia.edu/pages_html/page_mission.html. Accessed 18 June 2015

Krauss L (2012) A universe from nothing. Free Press, New York, p 25

Krauss L, Scherrer RJ (2007) The return of a static universe and the end of cosmology. Gen Relat Gravit 39:1545–1550

Liddle A (2003) An introduction to modern cosmology. Wiley, New York

Mayo D (1996) Error and the growth of experimental knowledge. The Chicago University Press, Chicago

Perlmutter et al (1999) Measurements of Ω and Λ from 42 high-redshift supernovae. Astrophys J 517:565–586

Popper KR (2002) The logic of scientific discovery. Routledge, London

Riess et al (1998) Observational evidence from supernovae for an accelerating universe and a cosmological constant. Astron J 116:1009–1038

Staley K (2011) The evidence for the top quark. Cambridge University Press, Cambridge

Susskind L (2006) The cosmic landscape. Back Bay Books, Boston

Susskind L (2007) The anthropic landscape of string theory. In: Carr B (ed) Universe or multiverse. Cambridge University Press, Cambridge, pp 247–266

Weber M (2004) The vocation lectures (edited by D. Owen and T.B. Strong). Hackett Publishing, Indianapolis

Weber M (2011 [1917]) Wissenschaft als Beruf. Duncker & Humblot, Berlin

Cosmology and Theology

Some Mistakes in the Cosmological Case Against God

Fabrice Pataut

Abstract

Cosmological results yield philosophical conclusions only with the help of extra-philosophical premises that must be grounded on non-question-begging arguments. I argue that in the cosmology case, the argument from cosmology to anti-theism is such that the philosophical juice would have to be supplied by an inference to the best explanation whose role is to justify the thesis that one may dispense with God in explanations of the existence of the universe. Problems abound with both in a case of extra-disciplinary transgression. My conclusion is accordingly negative: No genuine inference is provided here and no dispensability thesis offered an adequate ground. I argue that no additional premise may provide a bona fide argument with (i) either a scientific principle or some cosmological data among the premises, and (ii) an anti-theistic philosophical claim as a conclusion. The relevant details of converse arguments in favor of

I have benefited enormously from a stimulating paper by Genco Guralp (Department of Philosophy, Johns Hopkins University) for the writing of this paper. Guralp's "Cosmology and the End of Weberian Science" (included in this volume) was presented at the International Graduate Summer School on *Scientific Knowledge and the Transgression of Boundaries* held at the University of the Basque Country (UPV/EHU), Spain, from August 27 to August 31, 2012, in collaboration with the Institute for Technology Assessment and Systems Analysis of the Karlsruhe Institute of Technology (KIT-ITAS). I was his assigned discussant on this particular occasion. I have also benefited from discussions on related topics involving lecturers and participants of the International Graduate Summer School.

F. Pataut (✉)
Paris, France
e-mail: Fabrice.Pataut@univ-paris1.fr

© Springer Fachmedien Wiesbaden GmbH 2016   187
B.-J. Krings et al. (eds.), *Scientific Knowledge and the Transgression of Boundaries,* Technikzukünfte, Wissenschaft und Gesellschaft / Futures of Technology, Science and Society, DOI 10.1007/978-3-658-14449-4_8

indispensability theses based on inferences to the best explanation are taken into consideration, and so are stronger reasons, put forward by Immanuel Kant and Gottlob Frege, to reject the theistic claim within the confines of philosophy. Concluding remarks are offered to the effect that a genuine philosophical challenge emerges from Lawrence Krauss' and Robert Scherrer's claim regarding the return to a static universe. If they are right, we know now that we would have held in the past the false belief that the universe is static and that we would be holding the very same false belief in the future, when we know, although only as a matter of sheer luck, that it is indeed expanding.

Keywords

Cosmology · Dispensability arguments · God (existence of) · Inference to the best explanation · Krauss (Lawrence) · Theology

Suppose that we are able to explain the origin and existence of the universe without invoking God, by giving some scientific explanation. Lawrence Krauss, for one, claims that current cosmology is able to do that: Quantum processes combined with inflation and dark energy can explain how a universe may come into existence from nothing, so that any appeal to God in the explanation is deemed superfluous (Krauss 2012).[1] Is this a case of warranted extra-disciplinary transgression? Are cosmologists now allowed to debate with theologians on the question of God's existence and creation *ex nihilo* and with philosophers on the question of nothingness? Krauss and others such as Stephen Hawking (arguing from quantum cosmology) and Leonard Susskind (arguing from the anthropic principle) believe that they must enter the debate because advances in cosmology ground their theological pronouncements to the theologian's disadvantage and their philosophical pronouncements to the philosopher's disadvantage. Krauss, Hawking and Susskind thus argue *directly* from cosmology to theology and philosophy, as if their scientific jurisdiction licensed them the extra-disciplinary transgression (see Hawking and Mlodinow 2011; Krauss 2012; Susskind 2005).

Or so it seems. The very idea that they are indeed arguing in this particular way is, I shall argue, a misrepresentation of what they do. Cosmological results yield theological and philosophical conclusions, if any, *only with the help of extra hidden philosophical premises*. Such premises must of course be grounded on

---

[1]Note that the conception of God as prime mover or creator doesn't necessarily involve a further belief in the perfection of His creation or in divine revelation. In what follows, I shall not be concerned with either theodicy or revealed religion.

non-question-begging independent arguments. This is the price one has to pay for such theological and philosophical conclusions.

The point I wish to start with is that cosmologists, in these particular instances (quantum cosmology, the anthropic principle), must argue with the help of what Paul Benacerraf, examining the philosophical use of metamathematical results, has called "Princess Margaret Premises".

Let me briefly go back to the parable of the Cohens and Princess Margaret as told by Benacerraf to make the point clear (see Benacerraf 1996, pp. 9–10). When the Cohens accept—somewhat reluctantly—the goy Princess Margaret as the right girl for their son Abie on the basis of the staggering advantages their future grandchildren would benefit from (being heirs to the throne of England, etc.), only *half* the shatchen's job is done.[2] Abie still has to marry Princess Margaret. Without the Cohens' acceptance of the marriage broker's final offer, no result could be obtained, but more is needed nevertheless for the job to be rounded off. The Cohens' reluctant acceptance after turning down so many proposals is what Benacerraf mischievously dubs "the easy part"; what is needed now is the extra move which will bring the broker's efforts to their expected conclusion.

Just as "authors who brandish metamathematics as the authority for their philosophical conclusions are not always 'up front' with the particular [Princess Margaret Premises] to which, if pressed, they would have recourse" (Benacerraf 1996, p. 43) to justify these conclusions, authors who resort to cosmology as the authority for their theological and philosophical conclusions are not always "up front" with the extra premises they would hold true and should be ready to defend on independent grounds to justify their conclusions, if urged to provide some clarification.

In the metamathematical result case, Benacerraf remarks—*pace* John Lucas (see Lucas 1961) and Roger Penrose (see Penrose 1989)—that in order to argue from Kurt Gödel's first incompleteness theorem to the philosophical claim that we are *not* machines, some supply of philosophical "juice" must be provided.[3] In other words, some extra *independent* Princess Margaret Premise is needed to obtain the anti-mechanistic conclusion that our cognitive abilities transcend those of any machine. The extra premise to be supplied on independent grounds is that human

---

[2]A shatchen is a Jewish professional marriage broker.

[3]Gödel's first incompleteness theorem may be very roughly stated as follows: In any consistent formal system S within which a certain amount of arithmetic may be carried out, there are statements of S which may neither be proved nor refuted in S. Gödel's proof of the theorem mechanically produces a sentence—often called "the Gödel sentence" of S—which is indeed undecidable in S, i.e., neither provable nor refutable in S.

mathematicians are able to find the Gödel number of any theorem proving machine and, given that, prove its Gödel sentence, something *no* machine is able to achieve (see Benacerraf 1996, p. 31).

I shall argue that in the cosmology case, the allegedly direct argument from cosmology to anti-theism is formulated in such a way that the philosophical juice is to be supplied—*if it can be supplied at all*—by an inference to the best explanation, whose role is to justify a dispensability thesis to the effect that one may do away with God in explanations of the existence and origin of the universe. Problems abound with both, especially in a case of extra-disciplinary transgression, and my conclusion will be negative on both counts: No genuine inference to the best explanation is provided here and no dispensability thesis thereby offered an adequate ground.

Two points, I think, should be made. The first is that cosmologists who would be mistakenly granted such an inference and such a thesis would de facto *not* be arguing *directly* from cosmology to theology. The second is that, given the usual and indeed prominent formulation of the anti-theistic pronouncement, the most relevant or likely candidates to the role of Princess Margaret Premises (PMPs) both fall short of the task. One cannot embark on the philosophical voyage from cosmology to anti-theism with such PMP baggage. Not only may one not argue directly from cosmology to theology, but one may not argue by relying on an inference to the best explanation and a dispensability thesis. This is not to say that there may not be some *other* appropriate PMP baggage for the voyage, but I think it will be clear from what follows that one may have serious doubts about any such possibility in a situation of *extra*-disciplinary transgression of the kind we are considering here.

The crucial point of the anti-theistic argument is that since God plays no theoretical role in our best cosmological explanation of the origin and existence of the universe, we must dispense with God. Let us focus on the consequent of the conditional. Suppose that, as a matter of general rule, the mere fact that we *can* do something implies that we *must* do it, so that if we can dispense with God, we should proceed without further ado. The generality of the principle or recommendation is far from obvious. The converse certainly holds: We must be able to do whatever is prescribed or deemed preferable, and the particular case where elimination pure and simple is desirable may even seem paradigmatic. But the inference from *can* to *must* is quite different because it *presupposes* that bringing about X by way of an effective disposal thereby engenders a much preferable state of affairs Y—in our case a desirable theoretical state of affairs in which we have cured ourselves of any recourse to something we never needed in the first place, i.e., the recourse to a prime mover, to a creator of the universe. In other words, we have assumed that theism—or perhaps even creationism, through a literal reading of *Genesis* I 1–19—is the option one chooses by default when no alternative cosmological explanation is available. If we grant the presupposition that we are better off, cosmologically speaking, without the problematic and incautious existence claim, it looks that, rather than a bona fide argument, we get at best an overkill and at worst a rather trite observation. Either one is providing an obliteration of the target with more force than required, as if no less than quantum cosmology or the anthropic principle were needed to dispose of God, or one is merely pointing to the uncontroversial fact that God appears nowhere in contemporary cosmological explanations: Neither in the theory of quantum processes favored by Krauss and Hawking, nor in the theory of vacuum structures favored by Susskind.

In other words, under such an unsatisfactory formulation of the anti-theistic "argument", we are left with the uncomfortable impression that *if* the referent of the word "God", or of any suitably co-referential expression such as "prime mover", did (as it were *per impossibile*) play a theoretical role in our best cosmological explanation, it would be easy to find a way of reformulating that (allegedly) best explanation so that God (or a prime mover) would not play that role at all, and so that, as a consequence, the formulation of the explanation would not have to include an occurrence of the word "God" (or of any suitably co-referential expression such as "prime mover"). On the other hand, it is not *even* the case that we have some easy homework to do and that we may enjoy the excess of overkill by way of a substantiated dispensability thesis because "God" *is not* part of the vocabulary of the best cosmological explanation of the origin and existence of the universe, nor is any other suitably co-referential expression such as "prime mover".

Cosmologists who claim, either explicitly or implicitly, that they have a *direct* argument against theism are particularly unclear on this point. They may be claiming that something *within cosmology* plays the very same role that God plays in a theological account of the origins of the universe, but does a much better job (and indeed *the* best job) because it can explain creation out of nothing in a scientifically respectable way. Hawking and Mlodinow, for instance, remark that since quantum cosmology can explain the existence of many universes naturalistically, "[t]heir creation does not require the intervention of some supernatural being or god" (Hawking and Mlodinow 2011, pp. 8–9). The parallel is misconceived: Theology is not in any sense putting forward an explanation of how the universe "arise[s] naturally from physical law" or, for that matter, from any natural law—except perhaps in the derived sense that God is the free creator of natural laws, in which case we are back to square one since laws of nature would then result from a supernatural intervention. In any event, the universe, conceived as God's creation, *is not* the object of a prediction justified by anything one could invest with the name of "theological science". (The very same remark applies to god with the dismissive lower case "g" in its name—or to some particular divinity of any given polytheistic system—and a set of more than one created universe, i.e., a multiverse.) In other words, *nothing* in cosmology plays the very same role that the prime mover plays in theology. So, a fortiori, no inference to the best cosmological explanation—or to a better one than the theological or philosophical explanation—may come into play that would serve as a ground on which to justify the claim that the theoretical role God plays in a theological explanation is a dispensable one. There is nothing to dispense with in the first place because we are *not* in a situation where a theory $T1$ gives a better explanation than a theory $T2$. We are in a situation where $T1$ gives an explanation of the origin and existence of the universe from nothing in terms of natural law, and $T2$ gives a non-naturalistic account of the same thing in terms of divine intervention *ex nihilo* so that $T1$ and $T2$ are incommensurable with respect to explanation.[4] The point here is that the idea that naturalistic explanations are genuine explanations and that theological or philosophical "explanations" are pseudo-explanations, uninformative narrations and therefore no explanations at all, *precludes* the very possibility of a dispensability thesis being justified by an inference to a better explanation. "Better" misses the point: Since a naturalistic explanation such as the one favored by Hawking may not be compared to a theological account of creation couched in terms of divine intervention, there is indeed an unbridgeable gap in any "argument" from predictive power—as part of an inference to the best cosmological explanation—to the negative pronouncement with respect to God's existence.

Susskind, on the other hand, is careful enough to acknowledge a distinction in kind between the claim that philosophical conclusions may be derived from the

---

[4]We are neither in a case of intra-disciplinary transgression, nor in a case of interdisciplinary transgression, but, as remarked earlier, in a putative case of extra-disciplinary transgression. Guralp's three-fold distinction serves another purpose in his paper. He resorts to it to distinguish between (i) grand unified theories of elementary particles interactions being applied to cosmology, (ii) the implementation of Bayesian methods in cosmology, and (iii) the justification of philosophical and theological conclusions on the basis of cosmological results. I am using it here to stress that contemporary cosmology is neither (i) rejecting creation *ex nihilo* or divine intervention *from within* as an unworkable though genuine cosmological hypothesis, nor (ii) as a hypothesis coming *from another scientific discipline* that cannot be successfully integrated into current cosmology. In the extra-disciplinary case we are considering, the two disciplines (cosmology and theology, or cosmology and philosophy) do not even share theoretical terms so that the extra-disciplinary transgression may not be carried out.

anthropic principle and the cosmological data about vacuum structures when he remarks that

> [w]e *can argue* the philosophical merits of the anthropic principle but we *can't argue* with quantitative information about the number of vacua with each particular property such as the cosmological constant, Higgs mass or fine structure constant (Susskind 2007, p. 262, emphases added).

Why can we argue with the conclusion and not with the premises? Presumably because philosophical or theological claims and scientific claims, either positive or negative, are not on a par, at least as far as inference to the best cosmological explanation is concerned, and if they are not, then a fortiori nothing within cosmology may play the role that God as a prime mover plays in theology, and nothing in theology may play the role that the anthropic principle plays in cosmology. For the very same reason, it is equally pointless to remark—turning now to quantum cosmology—that

> [...] if the quantum properties of matter and radiation end up endowing even an infinitesimally small region of empty space with energy at very early times, this region can grow to be arbitrarily large and arbitrarily flat. When the inflation is over, one can end up with a universe full of stuff (matter and radiation), and the total Newtonian gravitational energy of that stuff will be as close as one can ever imagine to zero [...] A universe from Nothing [...] indeed (Krauss 2012, p. 104).

Notice just how far we are from creation *ex nihilo*, i.e., creation from nothing, creation from non-being ab initio. Quantum properties of matter and radiation do not and indeed cannot amount to nothing; this is not because one would define "nothing" as one chooses, *à la* Humpty-Dumpty, so that "nothing" ends up never being "any of the versions of nothing that scientists currently describe" (Krauss 2012, pp. xiii–xiv). Krauss's irritated complaint is inappropriate. The point is that *any* descriptive version of nothing turns nothing into *something*.

Because of the independence and autonomy of the disciplines, no PMPs are available that would help us to work out a bona fide argument with a scientific principle such as the anthropic principle, or some relevant quantitative information, among the premises, and an anti-theistic philosophical or theological claim as a conclusion.

It will nevertheless be instructive to consider some specific details of the failure. One way to do this is to consider *converse* arguments in favor of *in*dispensability theses based on inferences to the best explanation. Advocates of such arguments hold that we must accept whatever objects we quantify over in a scientific discourse delivering a best explanation because of these objects' ineliminable theoretical role. Unless we are able to show how to dispense with such roles, we must, however reluctantly, accept the objects that play them. It is also remarked in connection with the issue of theoretical role that these objects appear in the course of values of the variables bound by the objectual existential quantifiers occurring in the language of the regimented theory offering the best explanation, so that, unless we are able to provide an alternative formulation of the regimented theory, we must accept them in our ontology, qua values of the bound variables. The epistemological issue of theoretical role and the logico-linguistic issue of the binding of variables connect in the following way: Bound variables are part of the apparatus which exhibits the logical form of the sentences of the theory we must hold to be true, but it is the theoretical role of these values that takes responsibility for the ontological commitment.

A difficulty emerges from the idea that our ontological commitments are sensitive to considerations of indispensability and explanatory force in the case of the (alleged) undermining of the theological account of the origin of the universe. According to this conception of the way arguments pro and con theism interact, the view that God is the transcendent creator of the universe is inextricably linked to the view that God is part of our ontology. But God, for the theist, is not in any way part of our ontology. The prime mover may not be counted as part of "the furniture of the world" (to use a familiar phrase). The God of theism is indeed transcendent to the world and to whatever "furniture" it may "contain", in particular to what cosmology claims it contains (e.g., dark energy). One may *not* use Quine's criterion of ontological commitment according to which to be is to be the value of a bound variable *and* argue that there are "modes of existence" to solve the difficulty: Immanent existence for whatever is needed for the cosmological account of the origin and existence of the universe, and transcendent existence for theology's prime mover. The distinction between what belongs to the natural world and what is transcendent to it is invidious indeed as far as the criterion is concerned. So we are in a situation exhibiting a very strong disparity between the cosmology case and the theology case precisely because it is believed (mistakenly) that the controversial existence claim may be refuted by way of an inference to the best scientific explanation. The strong disparity emerging from the contention that a best explanation grounds an indispensability claim with respect to, say, dark energy and, by parity, a dispensability claim with respect to God as a prime mover working *ex nihilo*, amounts to this: Matter and radiation are *both* part of the ontology of cosmology *and* of the furniture of the world just in case cosmology says truly that the world came to existence because of an inflation, whereas God is part of the ontology of theology but *not* part of the furniture of the world just in case theology says truly that a transcendent God created the world *ex nihilo*.

One may wish to insist at this point that the prospect of proving the existence of a transcendent God has weakened considerably from within the confines of philosophy anyway, quite independently of cosmological results established within the confines of natural science, namely because of Kantian considerations firmly rooted in philosophical ground, so that cosmological considerations of a philosophical nature—as opposed to the scientific cosmological considerations we have been considering so far—are now ruled out from the start as far as the existence of God is concerned.[5]

Immanuel Kant's crucial point against the argument for the unconditioned existence of a first cause of the origin of the universe is that it relies on the ontological argument, which must be rejected (Kant 1965 [1781], B634). The Kantian attack on the cosmological argument builds on the idea that the conclusion to the effect that a necessary prime mover exists must be construed as the claim that a prime mover whose non-existence is impossible (in the sense that the denial of its existence is "absolutely inconceivable") exists as a matter of logical necessity (Kant 1965 [1781], B621). The ontological argument's basic contention is that we would be contradicting ourselves in denying the existence of God. Nothing, Kant remarks in this respect, may be determined to exist by virtue of its very concept involving necessary existence. In other words, no being is logically necessary, so that "God is a necessary being" may *not* be read or understood as meaning "The proposition 'God exists' is logically necessary". Kant's insight that existence is not a "determining" predicate on a par with other non-logical predicates is fully articulated in Gottlob Frege's claim that existence is a second-level predicate that may be assigned to first-level concepts and not to individual objects, and thus not, e.g., to God. Existence is thus not a first-order property of individuals but a second-order property of concepts (see Frege 1950 [1884], § 53). Frege's distinction solves the problem at the heart of the cosmological argument rejected by Kant in a most radical way: The proposition "Anything that has property X also has

---

[5]Note nevertheless that Richard Swinburne has proposed an Inductive Cosmological Argument (Swinburne 1979) that appeals to an inference to the best explanation whose role is to increase the probability of the conclusion "God exists". Swinburne considers theism to be the best explanation of the complexity of the universe, ruling out humanism and materialism as unlikely candidates. "Explanation" in this context, directly appeals to God's intentions and actions. The idea that an explanation of the origin and existence of the universe is complete (and therefore the best) only insofar as the intentions and actions of a conscious agent are taken into consideration presupposes that God is the uncaused *cause* of the universe, acting *from without*. The Swinburnian theistic explanation, allegedly being better for not letting the existence of the universe amount to a brute fact (something Hawking is not guilty of), illustrates the disparity just mentioned.

the property of existing" may not be expressed at all because existence is not a property we assign to objects. So we are not, despite appearances, ascribing the real property of existence to the unique object falling under the concept GOD when we claim that God exists. Existence is not, *pace* Saint Anselm, an *additional* property we attribute to God, so that a being having all of God's properties minus existence would be less perfect than Him.

Although God does play a theoretical role, qua prime mover, as the referent of the word "God" in a theological account of the origin and existence of the universe, the referent of the expression "being whose non-existence is impossible" may not play that role, because of the deductive failure diagnosed by Kant. Kant's insight and Frege's construal of existence as a second-order property of concepts thus seem to offer strong reasons to conclude that God, conceived as a being whose existence it would be contradictory to deny, may not as a matter of fact play *any* part in a theological explanation of the existence and origin of the universe. So even if, as it were *per impossibile*, one could argue from scientific cosmological results and principles to a denial of God's existence with the PMP baggage we have previously considered (and rejected as inefficient), the negative conclusion to be justified, i.e., "God does not exist" couldn't be cashed out as the denial that an uncaused cause of the origin and existence of the universe, i.e., a prime mover, necessarily exists. It seems that there are *stronger* reasons to deny this claim so that the very idea of a justification of the interdisciplinary transgression via a dispensability claim applied to the particular case of God as prime mover is threatened by purely philosophical considerations of a Kantian nature. No matter how hard we squeeze, we don't seem to get any philosophical or theological juice from current cosmology with the help of an inference to a best explanation and a dispensability claim as PMPs.

I now want to offer some concluding remarks concerning another argument related to the one we have considered so far. The argument is proposed by Krauss and Scherrer and is based on the remark that,

> [...] as we extrapolate the current ΛCDM universe forward in time, all evidence of the Hubble expansion will disappear, so that observers in our 'island universe' will be fundamentally incapable of determining the true nature of the universe, including the existence of the highly dominant vacuum energy, the existence of the CMBR, and the primordial origin of light elements. With these pillars of the modern Big Bang gone, this epoch will mark the end of cosmology and the return of a static universe (Krauss and Scherrer 2007, p. 1545).[6]

---

[6]In the ΛCDM cosmological model, Λ is the cosmological constant introduced by Einstein to allow the gravitational field equations to predict a stable universe and "CDM" stands for

Given the effects of the dark energy at the origin of the universe,

[o]bservers when the universe was an order of magnitude younger would not have been able to discern any effects of dark energy on the expansion, and observers when the universe is more than an order of magnitude older will be hard pressed to know that they live in an expanding universe at all, or that the expansion is dominated by dark energy. By the time the longest lived main sequence stars are nearing the end of their lives, for all intents and purposes, the universe will appear static, and all evidence that now forms the basis of our current understanding of cosmology will have disappeared (Krauss and Scherrer 2007, pp. 1549–1550).

The philosophical claim one is now grounding on cosmology is that we know *now* that (i) we would not have been able earlier to recover any dark energy related evidence to the effect that we are living in an expanding universe, and that (ii) we will not be able to recover that very same dark energy related evidence at a later time. In other words, we know at time t_0 that we would have falsely believed in the static nature of the universe at time t_{-1} and that we would falsely believe the very same thing in the future at t_{+1} had we not been lucky enough to gather at t_0 the evidence that we have indeed gathered. We know now that, just because the universe expands, future observers will not be able to receive light from galaxies of the universe, so that our best observational and theoretical tools give us reason to believe that although the universe will appear static to us, it will as a matter of fact continue to expand. This, it seems to me, provides a genuine philosophical puzzle that does not rely on any pernicious ambiguity in the use of the word "cosmology" in the first quote: The "end of cosmology" might mean either the very end of that scientific enterprise, along with the ontological commitments pertaining to it, or the end of the cosmological data and events that are still accessible today but will not be in the future due to the static appearance of the universe.

To know with that degree of certainty that nothing we may think of today may tell us how we might ground the cosmological beliefs we nevertheless know will be true in the future independently of anything that will be then available to us to justify such beliefs—*that* present knowledge of a lack of a basis for our future knowledge certainly is food for philosophical thought. One should indeed like to reveal the unexpected premises and consequences of such a philosophical position when it is informed by established cosmological results.

(Footnote 6 continued)
cold dark matter. Λ is interpreted in current astro-particle physics as referring to a vacuum energy density. "CMBR" stands for cosmic microwave background radiation. I am indebted to Guralp's paper for both quotation and explanation.

References

Benacerraf P (1996) What mathematical truth could not be - I. In: Morton A, Stich SP (eds) Benacerraf and his critics. Basil Blackwell, Oxford, pp 9–59

Frege G (1950 [1884]) Grundlagen der Arithmetik. Eine logisch-mathematische Untersuchung über den Begriff der Zahl/The foundations of arithmetic: a logico-mathematical enquiry into the concept of number. English translation by J. L. Austin with German text. Blackwell, Oxford

Hawking S, Mlodinow L (2011) The grand design. Bantam, New York

Kant I (1965 [1781]) Immanuel Kant's Critique of pure reason. (English trans: Kemp Smith N). St. Martin's, New York

Krauss LM (2012) A universe from nothing: why there is something rather than nothing, with an afterword by Richard Dawkins. Free Press, New York

Krauss LM, Scherrer RJ (2007) The return of a static universe and the end of cosmology. Gen Relativ Gravit 39:1545–1550

Lucas JR (1961) Minds, machines and Gödel. Philosophy 36:112–127

Penrose R (1989) The emperor's new mind: concerning computers, minds, and the laws of physics. Oxford University Press, Oxford

Susskind L (2005) The cosmic landscape: string theory and the illusion of intelligent design. Little, Brown and Company, New York

Susskind L (2007) The anthropic landscape of string theory. In: Carr B (ed) Universe or multiverse. Cambridge University Press, Cambridge, pp 247–266

Swinburne R (1979) The existence of God. Clarendon, Oxford

Part III

Radical Transgression of Boundaries

Boundaries Between Territories of Knowledge
Colonization or Independence?

Jaume Navarro

Abstract

In this paper, I argue that scientific disciplines are not natural kinds; rather their boundaries and limits are the result of contingent, historical processes. In his inductive philosophy, the Cambridge-based, influential polymath William Whewell depicted scientific fields as each referring to one definite object, based on one particular principle and largely independent from the rest of disciplines. This static portrait of the division of science contrasts with the history of the configuration of disciplines in the old English university during the second half of the nineteenth century. Taking the career of the physicist Joseph John Thomson as a case study, I describe the process by which physics became institutionally distinct from chemistry, in spite of his constant attempts to create a large department for what he called the "physical sciences", which would include physics, chemistry and engineering. Moreover, his interest in spiritualism strengthened his views on the unity of all science and the continuity between different "provinces of knowledge". Interestingly, this unity was instrumental in his work on electrical discharge in gases and the eventual discovery of the electron. Thus, I argue that the boundaries between disciplines should be readily transgressed, following the circulation of knowledge, methods and principles before such boundaries actually crystallized.

Keywords

William Whewell · Joseph John Thomson · Scientific disciplines · Spiritualism · Pluralism · Wave-particle duality

J. Navarro (✉)
Donostia-San Sebastián, Spain
e-mail: jaume.navarro@ehu.es

© Springer Fachmedien Wiesbaden GmbH 2016   201
B.-J. Krings et al. (eds.), *Scientific Knowledge and the Transgression of Boundaries,* Technikzukünfte, Wissenschaft und Gesellschaft / Futures of Technology, Science and Society, DOI 10.1007/978-3-658-14449-4_9

English nineteenth-century rhetoric of science is replete with metaphors related to imperial terminology. Knowledge is often depicted as a territory to be explored, fenced and colonized by adventurers at the service of the empire. In analogy to campaigns overseas, science would progress from the known to the unknown by charting territories (natural history), securing strategic points (inducing first principles), and extending a network of power (deducing consequences from such first principles). Furthermore, the chart of knowledge was to be split into clearly delimited fields or disciplines, based on their natural objects, similar to the division of overseas dominions on the grounds of geographical boundaries, natural resources or ancient tribal territories.

This rhetoric was highly present in the writings of William Whewell (1794–1866), a very influential figure in the reforms that took place in the division of knowledge in Cambridge in the early nineteenth century. Often referred to as a polymath, Whewell was a key figure in reforming the curricula and organization of knowledge in Cambridge and, from there, to Victorian Britain. His philosophy partly materialized in the implementation of the new "triposes", the exams that configured the pedagogical itineraries in the old university.[1]

Whewell's philosophy of science took *his* history of science as its foundational stone. *The History of the Inductive Sciences* was published in 1837, 3 years before he published the first edition of his *The Philosophy of the Inductive Sciences, Founded Upon Their History* (Whewell 1840), as a way to demonstrate that his inductive philosophy for the sciences was grounded on the very history of each and every science. Using imperial analogies, he warns the reader that in the task of developing his philosophy of science,

> I shall take, as a chart of the country before me, by which my course is to be guided, the scheme of the sciences which I was led to form by travelling over the history of each in order (Whewell 1840, vol. 1, p. 79).

Thus, the *History* serves the purpose of classifying the initially wild territory of knowledge.

In the first edition of the *Philosophy*, we find a table with the classification shown in Fig. 1. This "Classification of Sciences must result from the consideration of their nature and contents" which his *History* has revealed. His classification depends on "the Ideas which each science involves", where "Ideas regulate and connect the facts, and are the foundations of the reasoning, in each science" (Whewell 1840, vol. 2, p. 278). His inductive philosophy highlights the importance

---

[1]See especially Warwick (2003). On Whewell, see Yeo (1993), Fisch and Schaffer (1991), and Snyder (2006).

CLASSIFICATION OF SCIENCES. 281

Phenomena but their Causes; and generally, as in those cases, their Mechanical Causes.

Fundamental Ideas or Conceptions.	Sciences.	Classification.
Space	Geometry	Pure Mathematical Sciences.
Time		
Number	Arithmetic	
Sign	Algebra	
Limit	Differentials	
Motion	Pure Mechanism	Pure Motional Sciences.
	Formal Astronomy	
Cause		
Force		
Matter	Statics	Mechanical Sciences.
Inertia	Dynamics	
Fluid Pressure	Hydrostatics	
	Hydrodynamics	
	Physical Astronomy	
Outness		
Medium of Sensation	Acoustics	Secondary Mechanical Sciences. (Physics.)
Intensity of Qualities	Formal Optics	
Scales of Qualities	Physical Optics	
	Thermotics	
	Atmology	
Polarity	Electricity	Analytico-Mechanical Sciences. (Physics.)
	Magnetism	
	Galvanism	
Element (Composition)		
Chemical Affinity		
Substance (Atoms)	Chemistry	Analytical Science.
Symmetry	Crystallography	Analytico-Classificatory Sciences.
Likeness	Systematic Mineralogy	
Degrees of Likeness	Systematic Botany	Classificatory Sciences.
	Systematic Zoology	
Natural Affinity	Comparative Anatomy	
(Vital Powers)		
Assimilation		
Irritability		
(Organization)	Biology	Organical Sciences.
Final Cause		
Instinct		
Emotion	Psychology	
Thought		
Historical Causation	Geology	Palætiological Sciences.
	Distribution of Plants and Animals	
	Glossology	
	Ethnography	
First Cause	Natural Theology.	

Fig. 1 The classification of sciences in William Whewell's *the philosophy of inductive sciences* (Whewell 1840, vol. 2, p. 281)

of the foundational ideas of each science, ideas that constitute the basis for the further development of every field. Indeed,

> [...] their progress depends upon the distinctness of certain fundamental ideas; and [...] these ideas being first clearly brought into view by the genius of great discoverers, become afterwards the inheritance of all who thoroughly acquire the knowledge which is thus made accessible (Whewell 1838, p. 20).

Thus, every single science, every discipline, is delimited and distinct from any other due to its foundation on one particular idea.

Just to mention a few, geometry would be founded on the idea of space, dynamics on matter, chemistry on affinity, biology on organization or psychology on emotion. The point I want to stress here is that, for Whewell, as for many other philosophers and practitioners of science after him, the sciences are *naturally* organized in separate fields of study. Certainly, many contemporary historians of science have shown that the emergence of particular fields or disciplines is always a historically contingent process. But even in that case, there seems to be a naturalization after the event; i.e., an implicit endorsement that the separation between scientific fields is here to stay and that all we can aspire to is attempts of inter-, trans- or cross-disciplinarity.

Drawing from my expertise on late-nineteenth-century British science, I want to argue that the history of science can shed light on the ways historical actors have tried to build bridges between separate fields of study in the past or, on the other hand, how certain fields have split and turned into separate disciplines. My case study will be the early career of Joseph John (J.J.) Thomson (1856–1940), a Cambridge physicist, whose career partly shows the frustrations generated by an increasingly divided territory of the sciences and the generation of more and more specialized fields, disciplines, and departments.

In the first section, I shall consider Thomson's attitude towards the distinction between physics and chemistry and his (unsuccessful) political moves to create a large department of the *Physical Sciences* in the University of Cambridge that would include both physics and chemistry. As we shall see, his attempts to erase the institutional gap between these neighboring fields of knowledge were guided by the idea that physics was the science that had achieved maturity and which other sciences had to imitate. Chemistry would, in this framework, be on the right track to become an adult science and, therefore, potentially subsumed into the broader category of *The Physical Sciences*. This territorial imperialism of physics was also present in the blurred limits between materialism and spiritualism that we find in Thomson's career and that of many of his contemporaries.

In the second section, we shall come across the *status quo* of spiritualism as a legitimate field of study within the *physical sciences* and not as a topic in the fringes of science only fit for mystics and charlatans. The episode will reveal that keeping a dogmatic limit between what may and may not be the subject of scientific study is not necessarily the most scientific attitude. Although the line taken by Victorian physicists did not take them very far, it certainly did encourage attention to some of the phenomena treated by early psychology.

In the third section, we shall step onto the metaphysics of matter. Thomson's career took place at the time quantum physics was emerging. Classical notions of causality, rationality or observation, just to mention a few, were seriously challenged and reformulated. If we follow a Kuhnian pattern, the old worldview had to be abandoned in order to embrace the new paradigm. We shall see that those scientists of the old generation who, like Thomson, refused to accept the quantum worldview did so not in spite of the evidence against the old physics but in order to preserve their epistemology. In this third section, I shall argue for pluralism in the sciences and the benefits of preserving old explanatory tools in modern science.

1 Physics, Chemistry and the Physical Sciences

A tourist visiting Cambridge will most likely be shown the old building of the Cavendish Laboratory on Free School Lane. The guide will talk about James Maxwell and the early days of the laboratory and then turn to two plaques on the façade: One commemorating the discovery of the electron in 1897 by J.J. Thomson, the other the elucidation of the helix structure of DNA by James Watson and Francis Crick in 1953. The knowledgeable tourist may ask why it was that DNA, a biological structure, was unraveled in a physics laboratory and not in the biology department, to which the guide will probably answer that the structure of DNA was the outcome of a long tradition in crystallography, thus cancelling the disciplinary doubts of the visitor. It is unlikely, however, that anybody questions the legitimacy of the Cavendish to host the discovery of the electron since the first elementary particle is, *obviously*, a topic for physicists.

In this section, we shall have a look at the career of J.J. Thomson leading to his finding of electrons and argue that his project was not an exclusively physical one but included much of what people at the time considered to belong to the chemical realm. In preparation, however, let us continue the tour on Free School Lane. At the end of the street, the guide may point at the building on the corner with Downing Street where the old Chemistry Department lies. Built in 1888, more than a decade after the inauguration of the Cavendish, this building symbolized the separation between physics and chemistry in the architecture of the university, destroying the dreams of Thomson and others to create a big department of *physical sciences* that would include chemistry. In 1894, his hope for merging both fields of research under the same umbrella was not over. In his lecture at the British Association for the Advancement of Science he used the following metaphor:

> The work of chemists and physicists may be compared to that of two sets of engineers boring a tunnel from opposite ends—they have not yet met, but they have got so near that they can hear the sounds of each other's advances (Thomson 1894, p. 493).

If we follow this metaphor in our touristic visit, we might be tempted to explore the space between both buildings, only to find out that the dreamt-of connection between both departments was filled, in the early twentieth century, not by a tunnel or a passage-way but by a third department: That of physical chemistry. Specialization won the battle against those hoping for more inclusive departmental umbrellas.

In 1909, Thomson would complain that the search for the unity of science was threatened by the ongoing specialization of young students, which

> [...] injures the student by depriving him of adequate literary culture, while when it extends, as it often does, to specialization in one or two branches of science, it retards the progress of science by tending to isolate one science from another. The boundaries between the sciences are arbitrary, and tend to disappear as science progresses. The principles of one science often find most striking and suggestive illustrations in the phenomena of another (Thomson 1910, p. 4).

This view of the arbitrariness of boundaries was manifested in the way that he directed the Cavendish in the first years of his tenure.

To what extent did this process relate to Whewell's earlier ideas on the essential distinction between areas of knowledge? And how did Thomson understand this separation? To answer these questions we need to take into account that, in Victorian Cambridge, the big separation was between mathematics and the natural sciences, and that most parts of physics were included in the former. Together with territorial metaphors, the other image that Whewell used was that of adult and under-aged sciences. The former would be those that had succeeded in the inductive process of finding their first principles from which to deduce all related phenomena. The latter were those sciences still engaged in their inductive process, i.e., in finding the very first principles from which to work mathematically or logically. Certainly, chemistry was still in the under-age stage of its development, while most physical topics had already achieved maturity: The last to do so, electricity and magnetism through the work of Maxwell. Thus, when Thomson was hoping to embrace chemistry within the big umbrella of the *physical sciences*, he was not saying chemistry was essentially the same but that it was in the verge of acquiring adulthood by the formulation of chemical principles in mechanical and dynamical terms, as Maxwell had done with electromagnetism.

Let us have a look at the steps J.J. Thomson took in his early career that may help us illustrate his Unitarian idea of the *physical sciences*. Graduating as second wrangler in 1880 in the Cambridge Mathematical Tripos, Thomson was, in principle, the perfect product of the wrangler system: One that trained physicists as applied mathematicians, never having any contact with experimental science. His skills as an applied mathematician were revealed when he won, in 1882, the prestigious Adams Prize. The subject for that year was "an investigation of the action of two vortex rings on each other", and it was typical of the Cambridge of the day. The question of vortex rings had been present among mathematicians and physicists since 1867, when William Thomson had suggested an atomic model in which atoms could be represented as vortex rings in the ether. Certainly, by 1882 W. Thomson had given up this cosmological idea, but the topic remained of interest to mathematicians. Most of the papers on vortex theory were published in journals of mathematics, not physical journals, because it was regarded as a most interesting mathematical problem. Not only hydrodynamics was involved, but also the new area of the topology of knots (Kragh 2002, p. 46).

In this situation, the 1882 Adams Prize was intended mainly as an exercise with a purely mathematical interest, but J.J. Thomson managed to broaden the question and to turn the problem of the stability of two vortex rings into an all-embracing theory of matter, thus giving a revival to the group of Victorian physicists who were involved in developing the idea of vortex atoms. In his words, an atomic theory based on the behavior of vortex rings

> [...] has á priori very strong recommendations in its favour. For the vortex ring obviously possesses many of the qualities which a molecule that is to form the basis of a dynamical theory of gases must possess. It is indestructible and indivisible; the strength of the vortex ring and the volume of liquid composing it remain for ever unaltered; and if any vortex ring be knotted, or if two vortex rings be linked together in any way, they will retain for ever the same kind of be-knottedness or linking. These properties seem to furnish us with good materials for explaining the durable qualities of the molecule (Thomson 1883a, p. 1).

"On Vortex Rings" reveals in its last section that all these calculations "would enable us to work out a complete dynamical theory of gases" (Thomson 1883a, p. 51). Atoms could thus be represented in terms of these vortex rings in a fluid, and he concluded that there could be stable combinations of up to six such rings. This is completely in agreement with the possible valences of most elements, and he was led to speculate as follows: "The atoms of the different chemical elements are made up of vortex rings all of the same strength, but some of these elements consist of only one of these rings, others of two of the rings linked together, others of three, and so on", which meant that "each vortex ring in the atom would correspond to a unit of affinity in the chemical theory of quantivalence" (Thomson 1883a, p. 54). In this model, the mass of the atoms ceases to be their fundamental characteristic and their chemical affinity assumes such a role. This shows that Thomson was very much interested in chemical combinations of elements and substances as a way to better understand the foundations and the constitution of matter.

The essay on vortex rings is, I think, revealing about the way in which Thomson approached chemistry. Physical chemistry was, at the time, an emerging field interested in explaining the physical mechanisms of the chemical processes; that is, in getting rid of affinities as some sort of force that was different from mechanical forces. That Thomson was trying to reduce chemistry to physics was evident to those who read the essay in Cambridge. G.H. (George Howard) Darwin, the professor of astronomy, would congratulate Thomson for winning the Adams Prize in the following terms:

> The problems you have solved are of amazing difficulty, and the results of the greatest interest. May you go on and discover a true dynamical theory of chemistry.[2]

Nevertheless this aim was not exclusive to Thomson. In 1885, for instance, G.F. (George Francis) FitzGerald would correspond with Thomson while trying to develop a model of the electromagnetic ether, saying:

> I thought it possible that electrical forces might be explained by these general effects of vortices & c. and that chemical forces might be due partly to these and partly to actions produced by the distortions of the vortices. For though chemical and electrical forces are due to like causes nevertheless chemical action is of a much higher order of complexity than simpler electrical actions.[3]

With this essay as his only major work, Thomson was appointed professor of experimental physics and director of the Cavendish Laboratory in 1884. By then, not only was he a young and quite inexperienced second wrangler, but the laboratory also was a recently created institution with no clear role in the University of Cambridge.[4] The origins of the Cavendish Laboratory show a tension in the role of

---

[2]G.H. Darwin to J.J. Thomson, 25 January 1883, Cambridge University Library Manuscripts (CUL), Add. 7654, D4.

[3]G.F. FitzGerald to J.J. Thomson, 1 January 1885, CUL, Add. 7654, F15.

[4]For a history of the first years of the Cavendish Laboratory, see Kim (2002).

a university laboratory in Cambridge. While Germany and France had discovered the benefits of large-scale research centers that were closely related to the universities, Cambridge was perhaps one of the last major academic institutions in Britain to accept the new role of research in the context of a university (see Sviedrys 1976; Gooday 1990). Especially in Cambridge, where physics was understood in terms of mathematical expertise, the laboratory could only be understood as a place in which to incarnate such mathematical precision (see Schaffer 1992).

In the first years of Thomson's directorship, the laboratory saw a number of changes. Partly due to his lack of experience, but partly also because of a lack of any experimental tradition in Cambridge, the Cavendish developed a "laissez-faire" policy. It is commonplace to say that there was no such thing as a Cavendish school before 1895, but that it became a place in which individual scientists were allowed to develop their own projects. There are four reasons why this policy developed under Thomson's guidance: (i) his lack of manual skill; (ii) his fertile imagination which prevented him from confining himself to routine experiments; (iii) his preference for visual rather than metrical approximations; and (iv) his preference for qualitative rather than exact results (see Falconer 1989, p. 108).

This laissez-faire policy can also be understood, however, as the result of a particular idea of the way that the sciences evolve, and not as a consequence of the young Thomson's lack of a program. Proof of that is the fact that Thomson did have a project of his own, with both a theoretical and an experimental side to it. After his work for the Adams Prize, Thomson undertook a study of the phenomena of electrical discharge through gases, which eventually led him to the discovery of the electron. It is interesting that Thomson dedicated himself to such a topic: From an experimental point of view, it was the field mainly of amateur scientists and it did not involve the precision and accuracy needed for it to be thought of as a serious Cambridge science. This did not bother Thomson, for he never thought these were the most relevant characteristics of experimental science. He was not interested in quantitative precision, or in a close match between theory and experiment, being satisfied with a rough qualitative comparison (Falconer 1989, p. 107). But, from a theoretical point of view, he was convinced that the apparently catastrophic phenomena occurring in the gases in discharge tubes would prove to be relevant to an explanation of a very fundamental issue: That of the interaction of electricity with matter, and thus to a better understanding of the nature of both substances. As a matter of fact, Maxwell had written in his *Treatise* a decade earlier that

[…] these, and many other phenomena of electrical discharge, are exceedingly important, and when they are better understood they will probably throw great light on the nature of electricity as well as on the nature of gases and the pervading space. At present, however, they must be considered as outside the domain of the mathematical theory of electricity (Maxwell 1891, p. 61).

His research on the behavior of electric discharge in tubes had a physical as well as a chemical side to it. From his early days as a schoolboy in Manchester, Thomson was fascinated by the problems of chemical combination and chemical structure (see Chayut 1991, p. 528). Electrolysis was the model that he used to account for the phenomena he observed in the discharge tubes. The different mental mechanical models that he proposed in the period prior to 1897 had in common electrolysis as a heuristic tool. The major point was that molecules in the gas split to make the transfer of charge possible. Although the specific mental models that Thomson proposed changed with time, the splitting of the molecules was a permanent characteristic, and this was an important concept that eventually led to the speculation that electrons were the carriers of electricity in cathode rays. The importance given to electrolysis can also be traced back to another suggestion of Maxwell. In the *Treatise*, he stated that

[…] of all electrical phenomena electrolysis appears the most likely to furnish us with a real insight into the true nature of the electric current, because we find currents of ordinary matter and currents of electricity forming essential parts of the same phenomenon (Maxwell 1891, p. 374).

When Thomson abandoned the vortex atoms as an explanatory tool, he imagined a different ether-based structure, which he called Faraday tubes. With them he also tried to account for the affinity of elements and the formation of molecules, consistent with his idea of giving chemistry the status of a science dependent on physical models. In a paper of 1895, affinity is understood as a consequence of the way that Faraday tubes end in the atoms. The atoms are described in terms of gyroscopic structures and the moment of momentum created by the interaction of a Faraday tube with the vorticity and the momentum of the atom accounts for the existence of ions with a greater affinity to some atoms than to others. The gyroscopic structures would account for the electrochemical behavior of atoms, and would reappear in his atomic model in the early 1900s to explain the role of corpuscles in the atom. In this way, chemical behavior could be explained in terms of physical mechanisms and not simply as the result of electrical forces. As an example, Thomson points to the apparent asymmetry in the bond between hydrogen and chlorine: A negatively electrified hydrogen atom and a positive chlorine atom will experience less attractive force than a positive hydrogen atom and a negative chlorine atom, although the electrostatic force would be the same. This is what Thomson meant when he said that "when charged atoms are close together, there may be forces partly electrical, partly chemical, in their origin in addition to those expressed by the ordinary laws of electrostatics" (Thomson 1895, p. 518). In his 1904 book *Electricity and Matter*, a compilation of a series of lectures given at the University of Yale in 1903, Thomson regrets that even though the idea that the chemical forces are of electrical origin has many supporters (Berzelius, Davy, Faraday, Helmholtz), "chemists seem, however, to have made but little use of this idea, having apparently found the conception of 'bonds of affinity' more fruitful" (Thomson, 1904, p. 133).

A last example of his constant interest in chemical affinity can be taken from a long paper published in 1914, "Forces between Atoms and Chemical Affinity". In it, Thomson studies the electrical forces of neutral atoms, assuming that these can be understood as electrical doublets due to the arrangement and the mobility of the negatively charged corpuscles in the positive electricity of the atom. From this asymmetry, the forces between atoms in one molecule as well as the intermolecular forces can be accounted for. In this model, the mobility of the corpuscles is important, since an atom with mobile corpuscles is more flexible in arranging its charges and, thus, forming molecules with other atoms, than that with fixed corpuscles. The latter would be saturated atoms. In agreement with what is seen in Mendeleev's table, the number of free corpuscles in an atom ranges from 0 to 8; and "when the number reaches 8 the ring is so stable that the corpuscles are no longer mobile and the atom is so to speak self-saturated" (Thomson 1914, p. 781). The way to saturate an atom with free corpuscles is, in his view, the following. When an atom is free, the tubes of force of every corpuscle end in the positive charge of the atom. The mobile corpuscle "will not be fixed unless the tube of force at its end is anchored to something not in the atom, i.e. it must end on another atom" (Thomson 1914, p. 782). The valency of an atom is, thus, related to the tubes of force originating in the free corpuscles and going out of the atom.

Thomson's engagement with physics and chemistry as part of a unitarian goal had his reward in 1923. In a visit to the US organized by the Franklin Institute in Philadelphia, Thomson was received by Irving Langmuir, Gilbert Lewis and others as a founding father of physical chemistry. The reason for that was that by explaining electrical phenomena in terms of electrons, Thomson had triggered the study of the inner composition of the atom as well as the role this composition had in the intricacies of chemical bonding. The electron was, by that time, the patrimony of both physicists and chemists along the lines of Thomson's unitarian project.

2   Matter, Spirits and the Boundaries of the Sciences

> Ordinary material systems must be connected with invisible systems which possess
> mass whenever the material systems contain electrical charges. If we regard all matter
> as satisfying this condition we are led to the conclusion that the invisible universe—
> the ether—is to a large extent the workshop of the material universe, and that the
> phenomena of nature as we see them are fabrics woven in the looms of this unseen
> universe (Thomson 1907a, p. 21).

This is J.J. Thomson at the end of a public lecture in Manchester, in November
1907, in which he argued that the latest discoveries in electricity proved a deep
metaphysical unity in the world, one that stemmed from the real existence of the
ether and from the fact that all phenomena in Nature were a result of matter in
movement. Two analogies permeate the rhetoric of this lecture: The machine-like
fabric of the world, and the existence of an unseen universe that keeps that machine
in productive movement. The first comparison is only natural at the center of
British manufacturing industry. The latter resonates with a best-seller by Peter
Guthrie Tait and Balfour Stewart who had argued for a strong link between the
science of energy and the existence of spirits and their action in the visible
universe.

This mention of an *unseen universe* is not an isolated occurrence. In his
autobiography, J.J. felt the need to write one full chapter, albeit the shortest, on
psychic research, a topic to which he had devoted intellectual attention and
political support, especially through his membership of the Society of Psychical
Research (SPR), of which he was even vice-president for some time. His interest
on psychic research may be traced back to his youth, to his pre-Cambridge years.
The mixture of smoke and humidity that permeated the atmosphere in industrial
Manchester is a compelling image of one important aspect in the worldview of
Victorian scientists: The world of matter was equally permeated by an entity—the
ether—which was a major seat of energy and interactions, and the medium for the
transmission of light. The ether was supposed to be weightless but, at the same
time, rigid enough to transmit light waves. The question about the relationship
between ordinary matter and ether, between matter and energy, was at its speculative peak in the second half of the nineteenth century, giving the ether some
elements of science fiction among the educated public. Science made its existence
necessary; its characteristics made it open to mystery and to all manner of
speculation.

Balfour Stewart, a physicist in Manchester who happened to teach J.J. Thomson in junior school, used this cosmological idea to write, in 1875, together with the also physicist Peter Guthrie Tait, a bestseller on natural theology, called *The Unseen Universe* (Stewart and Tait 1875). Taking ether as the ultimate reality in Nature, Tait and Stewart tried to prove the immortality of the soul and the possible existence of many spiritual entities. The main idea was that the world as we know it, the "visible universe" as they put it, was only a minor part, contingent and finite in time, of a greater universe, the unseen universe, which would include all created things:

> We maintain that the visible universe—that is to say the universe of atoms—must have had its origin in time, and that while THE UNIVERSE is, in its widest sense, both eternal and infinite, the universe of atoms certainly cannot have existed from all eternity (Stewart and Tait 1875, p. 9).

In this context, the atoms of matter would be a transient entity:

> We are not led to assert the eternity of stuff or matter, for that would denote an unauthorised application to the invisible universe of the experimental law of the conservation of matter which belongs entirely to the present system of things (Stewart and Tait 1875, p. vii);

or, to put it more bluntly,

> [...] it appears no less false to pronounce eternal that aggregation we call the atom, than it would be to pronounce eternal that aggregation we call the sun (Stewart and Tait 1875, p. vi).

Matter was regarded as a non-fundamental entity in the complete universe, but only as an ephemeral phenomenon of the visible universe. Here they introduce a distinction between "objective" and "substantive" reality, saying that, while atoms have both types of reality, the unseen world of ether is "objective" but not "substantive", an idea that can only be understood in the light of the science of energy that crystallized in the previous decades:

> It is only within the last thirty or forty years that there has gradually dawned upon the minds of scientific men the conviction that there is something besides matter or stuff in the physical universe (Stewart and Tait 1875, p. 100).

And, continuing with the same kind of rhetoric, they take energy as this "something" besides matter:

> Taking as our 'system of bodies' the whole physical universe, we now see that [...] energy has as much claim to be regarded as an objective reality as matter itself (Stewart and Tait 1875, pp. 114–115).

In Tait and Stewart's views, there was, however, an ontological asymmetry between matter and ether, for the latter was considered to be more fundamental than the former. They regarded the ether as a non-perfect fundamental fluid, in which the vortices appear and disappear as a result of spontaneous fluctuations. In this way, the visible world would be ephemeral "just as the smoke-ring which we develop from air [...] is ephemeral, the only difference being in duration, these lasting only a few seconds, and the others it may be for billions of years" (Stewart and Tait 1875, p. 157).

This holistic idea was not at all characteristic only of Stewart and Tait: Late nineteenth century science was over-enthusiastic about the possibilities of reducing all knowledge to one metaphysical principle from which all phenomena, including the spiritual, would be deduced (see Harman 1982; Myers 1989; Smith 1998; Noakes 2005). *The Unseen Universe* is only one example of a "growing commitment to a belief in the uniformity of nature, the restriction of divine action to the creation of the universe, the rejection of suppositions of divine interventions to explain apparent discontinuities in the natural world, and the separation of the natural and the supernatural" (Heimann 1972, p. 75). The interest of this book, as far as J.J. Thomson is concerned, resides not only in the fact that it was a best-seller among those with interests in science and natural philosophy, but mainly in the fact that Stewart was writing this book precisely in the years when J.J. spent long hours in the laboratory under his guidance, and this must have certainly exerted a direct influence on him (Sinclair 1987, p. 90). As Davis and Falconer stated, Thomson received from Stewart a thorough grounding in the prevalent Victorian method of reasoning by analogy and in ether physics, and, perhaps also an interest in psychical research (Davis and Falconer 1997, p. 6; see also Chayut 1991).

This brings us back to the quote at the beginning of this section and Thomson's involvement in the Society for Psychical Research. The society, formally founded in 1882, was intended as a scientific response to the multitude of groups and associations interested in all kinds of paranormal phenomena in Victorian Britain. Interest in spiritualism and related issues had all sorts of motivations: As a way to prove the reality of an afterlife and the need for religion, as a means to challenge the authority of the established Anglican Church, as a response to the threat of increasing materialism, or as a way to extend the scientific ethos to the matters of the mind (Oppenheim 1985; Gauld 1968; Haynes 1982). The Society for Psychical Research was particularly cautious about the status of spiritual and psychic phenomena and its aim was "to examine without prejudice or prepossession and in a scientific spirit those faculties of man, real or supposed, which appear to be inexplicable in terms of any generalized hypotheses" (quoted in Gauld 1968, p. 137). The society was well respected among Cantabrigian academics since amongst its most enthusiastic driving forces were two Cambridge dons: Henry Sidgwick and Frederick Myers.

Sidgwick was a Trinity graduate in classics who became lecturer in moral philosophy and, eventually, Knightsbridge Professor of Philosophy in Cambridge. In 1869, he resigned from his appointment in Trinity College, since he no longer felt he could assent to the 39 articles of faith that fellows had to sign, although the College found extraordinary ways to keep him until 1882, when regulations were changed, and he was re-admitted as an ordinary fellow. Ever since his undergraduate years, Sidgwick had been involved in psychic research through the Cambridge Ghost Society, and, as a mature philosopher and Anglican apostate, he saw in spiritualism a possible way to support Christian morality without assent to its theological content (Oppenheim 1985, pp. 113–116). In J.J. Thomson's words, "he was one of the most brilliant talkers of his time [...] [and] the most brilliant in Cambridge" (Thomson 1936, p. 294), and greatly involved in reforms at the University. Particularly important was his and his wife's work in connection with women's education and the creation of Newnham College.

Sidgwick became the first president of the Society for Psychical Research. "He was an ideal president for such a society, absolutely fair and unbiased and critical" (Thomson 1936, p. 299): He was a highly respected and honest man who did not hide his many disappointments in his search for psychic and spiritual evidence. As a matter of fact, Sidgwick was "notoriously unlucky as a psychical researcher", and while people like William Crookes, Oliver Lodge or Alfred R. Wallace saw indubitable evidence of some paranormal phenomenon, Sidgwick, "in spite of repeated trials, [...] never witnessed anything" (Crookes, quoted in Oppenheim 1985, p. 124). His interest in the subject was challenged by many disappointments, turning him into a moderate agnostic and, thus, a respectable president of the society.

Myers was by far more enthusiastic than Sidgwick and the one who brought J.J. into the Society for Psychical Research: "in the nineties, at the instance of F.H.W. Myers, I attended a considerable number of séances at which abnormal physical effects were supposed to be produced" (Thomson 1936, p. 147). Although no longer a lecturer in Trinity College, he stayed in Cambridge as part of the intellectual elite, making himself a name as a poet, critic and essayist. He was first interested in psychic and paranormal research through the advice of Sidgwick, his undergraduate tutor at Trinity in the 1860s, but his interest had surpassed that of his tutor's from the 1870s onwards. Through the work of the Society for Psychical Research, he became increasingly convinced of the importance of hypnotism in developing a science of the psyche, and he developed the concept of the 'subliminal self', "the boldest and best known of the contributions that psychical research made to psychology before World War I" (Oppenheim 1985, p. 254).

In his recollections, J.J. seems to have mixed feelings about the activities of the Society for Psychical Research. He was clearly disappointed about the fact that "at all but two of those [séances] I attended nothing whatever happened, and in the two where something did there were very strong reasons for suspecting fraud" (Thomson 1936, p. 147). In spite of that, J.J., as well as Lord Rayleigh and some other physicists in Cambridge, "maintained a deep interest in the society's work but conducted only occasional investigations into psychical phenomena" (Noakes 2005, p. 426): They were observers and not active actors in this research. Thus, they had no serious grounds to dismiss an activity to which they were, at least in principle, not opposed. Their interest was probably not so much in psychology, let alone a belief in spirits and ghosts, but in the possibility of extending the domains of physics to the study of the mind. J.J. had a better attitude towards telepathy, "another branch of psychical research which may be connected with physics", and of which he had witnessed some positive instances. By the time he wrote his memoirs, J.J. still thought that "in my opinion the investigation of short-range thought transference is of the highest importance" (Thomson 1936, p. 154).

One highly popular story in Cambridge was the visit of an illiterate Italian peasant, Eusapia Palladino, who had acquired a name as a medium, in the summer of 1895. Invited by the Society for Psychical Research after Lodge and Myers were convinced of her powers at a séance in France (also attended by the more sceptic Sidgwick), Palladino performed a number of what many saw as deceptive tricks. Apparently, her behavior "stimulated the prejudices latent in the Sidgwick group", J.J. included. In a most ironic paragraph, J.J. also describes the case of the famous Madame Blavatsky in her visit to Cambridge:

One of my most interesting experiences was a séance when nothing at all happened [...] She said at the beginning that her Mahatma in Tibet would precipitate a message, a cushion and a bell, and we sat waiting for, I should think, more than an hour, and nothing whatever arrived. The medium was not in the least abashed. She took the offensive, said it was all our fault, that our scepticism had created an atmosphere impenetrable to anything spiritual. She was a short and stout woman with an amazingly strong personality, very able and an excellent speaker. So well did she speak that she convinced the great majority of the audience that the failure was their fault, and they went away thoroughly ashamed of themselves for having spoiled what would otherwise have been a most interesting experience (Thomson 1936, pp. 153–154).

In spite of this account, J.J. still thought that the Society for Psychical Research was a good idea:

> This work has not been wasted. To put its claims at the very lowest it is surely a great thing to have created an organisation for collecting and testing these abnormal phenomena and thereby to go far to ensure that no genuine ones will escape discovery (Thomson 1936, p. 299).

As Richard Noakes argued, Thomson's interest and membership in the Society for Psychical Research problematizes the positivistic view that, by the turn of the century, there was a clear-cut definition of the limits of physics. The history of psychic research needs to be seen "as an episode in late-classical physics" rather than as something alien to it (Noakes 2008, p. 326).

3 Corpuscles Versus Waves: A Problematic Distinction

What *is* a particle? Questions like this make many scientists and philosophers uneasy. The history of atomism and of fundamental particles is one with constant U-turns, dead-ends, sudden multiplication of entities and wonderful ideas to organize the chaos brought about by the inflation in the number of particles. More fundamental than the structure of matter is, actually, the quest on the nature of matter, which is not exactly the same. Thomson's career addressed both issues, although only the first is regarded as scientifically relevant. Thus, his discovery of the electron, portrayed as the first elementary particle in a long list to come, overshadows his stubbornness (from a modern perspective) to keep some sort of ether and, with it, a metaphysics of the continuum. But, as I have argued at length elsewhere, Thomson's career cannot be fully understood without the ether, even when it comes to explain his discovery of the electron (Navarro 2012). In this last section, I shall try to explain the ways Thomson tried to engage with material and energetic discreteness from a metaphysics of the continuum.

J.J. Thomson was interested in the constitution of matter and in linking mechanical, electrical and chemical phenomena within the same explanatory framework. In his first work, he unified electricity and mass, bringing forward the concept of electromagnetic mass (see Thomson 1881). Later, as we have already seen, the vortex theory presented him with an opportunity to unify mechanical and chemical phenomena. In his theory of Faraday tubes, starting in 1891, the three kinds of phenomena would be roughly united, an aim that he would continue to pursue after the discovery of the corpuscle and the various different atomic models he designed. The electron brought an apparently different model on the structure of the atom based on electrons. This section addresses all these models of matter with the following aim: That, at least for Thomson, a discrete model of the atom based on elementary discrete particles was not incompatible with a continuous idea of matter.

In April 1881, soon after his graduation, Thomson published a paper in the *Philosophical Magazine* in which he calculated the resistance of an electrified body in an electromagnetic field, a phenomenon that had been studied experimentally by Crookes and Eugen Goldstein. Thomson suggested that the behavior of high-speed charged particles could be explained by attributing to them an increase in apparent mass. His calculations assumed that

> [...] the resistance cannot be analogous to an ordinary frictional resistance, but must correspond to the resistance theoretically experienced by a solid in moving through a perfect fluid. In other words, it must be equivalent to an increase in the mass of the charged moving sphere, which we now proceed to calculate (Thomson 1881, p. 230).

In this way he introduced the idea of electromagnetic mass, which was going to be very relevant in the forthcoming decades.

The electromagnetic theory of matter that Thomson was tentatively suggesting was the first of a series of theories in which the ontological status of matter was a bit fuzzy. The observed mass was not a constant attribute of bodies, as a mechanical approach would suggest, but a variable magnitude. Following the hydrodynamic analogy, this also meant that the moving body had no definite spatial limits. The limits were imposed by the strength of the electromagnetic field and the speed of the particle. After the discovery of the corpuscle, he would even develop a model in which the only source of mass was the movement of charges, thus giving pre-eminence to continuity in the metaphysical explanation of matter.

In the vortex ring theory, the link between discrete and continuous comes from the fact that vortex rings are permanent (according to Helmholtz and Kelvin), so that atoms are always there, and no part of the ether ever becomes matter:

> Some portions of this [perfect fluid] are supposed to be rotating, the rest not: The rotating parts of the fluid on this theory are the atoms (Thomson 1914, p. 25).

Thus, the main characteristic of atoms (that they cannot be created nor destroyed) still holds. In 1914, while still making claims for the beauty of the vortex atom theory, Thomson would make it clear that the power of the theory resided in it being an ultimate explanation of matter, but not that any explanation of matter should be reduced to this approach. He would

[...] consider some of the properties of the individual atoms in this theory, remembering that if we took a collection of a large number of them, the properties of the aggregate would be those of ordinary matter (Thomson 1914, p. 25).

This is, I think, very significant, for there is no continuity of explanatory power. It is not that with the vortex ring theory any other approach to the material world is invalidated (the goal of extreme reductionism): They are valid approaches, but they are not ultimate explanations, which emphasizes the metaphysical character of the vortex theory.

With the vortex theory in mind, Thomson developed his first theory of dissociation to explain the discharges in high vacuum tubes. In June 1883, he published a paper in the *Philosophical Magazine* imagining the molecules of a gas dissociating due to an electric discharge in the gas (Thomson 1883b, p. 428). His program of studying the discharge of electricity in gases made it clear to him that there was a strong similitude between the discharge in gases and electrolysis in liquids. In 1890, this forced him to shift from an explanation of atomic bonds due to vorticity to an explanation based on electrical phenomena. Significantly, the importance he gave to the mechanisms of electrolysis to account for the discharge in gases enabled him to develop and work with the ideas of discrete matter and charge. The tension between discreteness and continuum had to be resolved both in the case of matter and of electricity. Maxwell's theory assumed a continuum displacement, not discrete charges. Thomson's new theory of Faraday tubes helped him to think of charge as a phenomenon at the end of them. He first achieved an atomization of charge in 1897 and of matter in 1899.[5]

But, what are Faraday tubes exactly? Electrostatically, they are unit tubes of electrostatic induction, all with the same strength corresponding to the electrolytic unit of charge. Mechanically, they are structures in the ether in the form of vortical tubes that begin and terminate in matter or form closed circuits. These tubes have a direction and the atoms on which they begin and end receive a unit of positive and negative electrification respectively. To support his new theoretical device, J.J. stressed that these structures were already present in the work of Faraday and

---

[5]See Falconer (1987), especially p. 243: "This attribute was not contained in Thomson's original corpuscle suggestion as is shown by his references in 1897 to it as a 'carrier' of electric charge; a continuation of his earlier ideas of charge as a surface interaction between Faraday tubes and material particles. He modified this view after 1899 as he worked out the corpuscular theory of matter, according to which the charge was the origin of the corpuscle's mass. [...] For Thomson there was no empty space and matter was simply a particular conformation of the all-pervading ether. Corpuscles were no exception. He envisaged corpuscles as located at the ends of tubes of electrostatic force which he thought of as vortices within the surrounding ether".

Maxwell, at least in the first sense of connecting positively and negatively electrified bodies (without the possibility of closed loops). In doing this, he set himself
in continuity with the two main British authorities on electricity and magnetism. As
a matter of fact, J.J. explicitly quoted Maxwell's description in the *Treatise* on how
to generate a tube of induction force from a line of force:

> If the line of force moves so that its beginning traces a closed curve on the positive
> surface, its end will trace a corresponding closed curve on the negative surface, and
> the line of force itself will generate a tubular surface called a tube of induction
> (Maxwell 1873, § 82).

As with the vortex-ring theory, Faraday tubes were structures emerging from the
dynamics of the ether. Since the ends of Faraday tubes had opposite electrifica-
tions, a tube had direction. Polarization in one point of a dielectric was a vector
quantity that represented the density and direction of Faraday tubes of force per
unit volume. Since Faraday tubes in a dielectric "cannot be created nor destroyed",
a change in polarization would take place only when tubes moved or deformed.
And, from these changes in polarization, which, at the end of the day, were
dynamical changes treatable with the usual methods of dynamics, Thomson
showed the direct connection between his method and the energy transfers
obtained using the Poynting vector.

Two atoms would form a stable molecule when united by a short tube of force,
i.e., a Faraday tube of molecular dimensions. At this stage, however, he did not
speculate much on the relationship between Faraday tubes and atoms of ordinary
matter. In his previous vortex theory, all phenomena were supposed to be mani-
festations of the ethereal vortex rings and their movements in the ether. There was
no asymmetry between ether and matter precisely because, in a way, matter did not
exist as essentially distinct from the ether but only as a manifestation of it. Furthermore, and more importantly, there was only one kind of structure in the ether:
The vortex rings. Now, on the other hand, atoms were essentially different from the
Faraday tubes, since these ended—or "fell", as he normally said—on atoms. It was
not until 1895 that he actually started to explicitly think on the relationship
between Faraday tubes, ether and atoms.

In the 1895 paper "On the Relation between the Atom and the Charge of
Electricity Carried by It", Thomson would talk about discrete electrical charges as
the ends in the ordinary matter of Faraday tubes. It is significant that when he
introduced the concept of discrete charge, this was accounted for in terms of a
continuous entity (the Faraday tubes in the ether). "Ordinary matter" was seen as
composed of gyrostats. The important thing is that the atom has a structure, a
mechanical structure, which gives it a particular behavior with the vorticity of the

Faraday tubes. Positive and negative ions will have different energies because their rotation is consistent with or contrary to that of the Faraday tubes.

In retrospect, the key moment of Thomson's career was his suggestion of the idea of the corpuscle. In my story this is also a significant issue, because it turns out that, in spite of his faith in the fundamentally continuous character of matter, Thomson is known as the person who found the first elementary discrete subatomic particle. However, the nature of the corpuscle was progressively modified. The 1897 corpuscle was a particle, with a ratio m/e 1000 times smaller than that of the hydrogen ion, which could be assumed to be present in all atoms. This could be deduced from the fact that the corpuscles were independent of the gas used in the cathode rays and that their mean free path was independent of their origin. These corpuscles were "matter in a new state, a state in which the subdivision of matter is carried very much further than in the ordinary gaseous state [...] a state in which all matter is of one and the same kind" (Thomson 1897, p. 312). Thus, the corpuscle brought a hint of hope in the attempts to unify matter, electricity and chemical properties, with only one fundamental component. The *only* problem was the atomization of matter to a more profound layer, a problem that could eventually be solved by considering the mass of the corpuscle to be of electromagnetic origin, thus reviving his hypothesis of 1881. This possibility was only mentioned in the 1899 paper, admitting that "we have no means yet of knowing whether or not the mass of the negative ion [i.e., the corpuscle] is of electrical origin" (Thomson 1899, p. 563). This idea would gain momentum in the following years and the continuous ether would be considered as the source of matter, electricity and the chemical properties of the atoms. In a trip to America in 1903, he would say:

> The view I wish to put before you is that it is not merely a part of a body which arises in this way, but that the whole mass of any body is just the mass of ether surrounding the body which is carried along by the Faraday tubes associated with the atoms of the body (Thomson 1904, p. 51).

As the existence of corpuscles was proposed, Thomson was aware of the importance of finding "the configurations of stable equilibrium for a number of equal particles acting on each other according to some law of force" (Thomson 1897, p. 313), a task that was manifestly impossible, due to the great number of equations involved, as well as the ignorance of such forces. However, he found the model of free floating magnets and their spontaneous configurations to be very suggestive, for such configurations kept proportions that resembled the relations in the periodic table. As seen in the examples in the previous section, a chemical property—the valence—was the characteristic that was used to determine the validity of a particular atomic model. Thus, the atomic model that evolved from the discovery of the corpuscle was a very simple one: One kind of element—the corpuscle—would be the only subatomic constituent. Positive electrification (certainly not positively charged particles), together with the mutual repulsions of the different corpuscles, would explain the arrangement of the corpuscles in atoms in a state of equilibrium (see Thomson 1903). Several thousand corpuscles in fast motion within the atom would account for its mass, since the positive electrification did not have any mass. He was still using his theory of Faraday tubes to explain why negative corpuscles had mass, while positive electricity didn't, saying that

> [...] the lines of force will therefore be very much more condensed near the corpuscle than at any other part of the system, and therefore the quantity of ether bound by the lines of force, the mass of which we regard as the mass of the system, will be much greater near the corpuscle than elsewhere (Thomson 1904, pp. 93–94).

Thus, positive electricity did not have any mass because Faraday tubes spread too much in the positive end, while they concentrated, forming the corpuscle, in the negative end. Again, the existence of discrete particles of matter was but a phenomenon of something more fundamental and of a continuous nature, i.e., the ether.

In the Victorian tradition, the ether is the entity that links mechanics and electromagnetism. In his book *Electricity and Matter*, written in 1903, Thomson would claim that the search for unity in the explanation of natural phenomena was closer to being completed than ever before:

> I have attempted to discuss the bearing of the recent advances made in Electrical Science on our views of the Constitution of Matter and the Nature of Electricity; two questions which are probably so intimately connected, that the solution of the one would supply that of the other (Thomson 1904, preface).

Even though he does not explicitly mention it in this book, the ether provides him with a continuous metaphysical entity, a characteristic which is important to avoid action at a distance:

> The fluid theories, from their very nature, imply the idea of action at a distance. This idea, although its convenience for mathematical analysis has made it acceptable to many mathematicians, is one which many of the greatest physicists have felt utterly unable to accept, and have devoted much thought and labor to replacing it by something involving mechanical continuity (Thomson 1904, p. 7).

Nevertheless, in 1906, Thomson realized that this simple model was not possible. His calculations of the number of corpuscles in atoms revealed this to be close to the atomic weight. This meant that corpuscles were only responsible for a minor part of the mass of atoms, and that the positively charged part of the atom accounted for most of the mass of the atom. This undermined the previous atomic model and triggered the search for corpuscles of positive electricity. Thomson would spend the following years trying to determine the nature of positive electricity in a similar way to that which had led him to the discovery of the negative corpuscle (see Falconer 1988).

The new model, which he started to develop in his book *The Corpuscular Theory of Matter*, would acknowledge that the search for ultimate unity was not at hand. The mass of the positive particles could not only be of electromagnetic origin, thus giving rise again to a dualism, with mechanical and electrical matter. His disappointment is clear at the beginning of the book, where he emphasizes that the new model has only a heuristic value:

> The theory of the constitution of matter which I propose to discuss in these lectures, is one which supposes that the various properties of matter may be regarded as arising from electrical effects. The basis of the theory is electricity, and its object is to construct a model atom, made up of specified arrangements of positive and negative electricity, which shall imitate as far as possible the properties of the real atom. The theory is not an ultimate one; its object is physical rather than metaphysical (Thomson 1907b, p. 1).

Not surprisingly, Thomson regarded the atomistic model of matter as purely *physical*, which, in his terms, meant that it was only a model. Continuity was absent in this model and, therefore, it could not be the last word as for the essence of matter.

Popular histories of J.J. Thomson regard his career after the discovery of the electron as irrelevant. I do not. Electrons became the highlight of his legacy, but his career continued with a constant reference to Faraday tubes, not to electrons. His experimental program shifted towards the study of what he called "positive rays". Theoretically, he was engaged, as many other physicists at the time, in the task to explain the increasingly contradictory properties of light and radiation, which appeared to behave at times as wave and at times as particles. Famous is his image, in the mid-twenties, of both conceptions being like "a battle between a tiger and a shark, each is supreme in his own element, but helpless in that of the other" (Thomson 1925, p. 15).

And it was in the attempts to explain the discreteness of radiation that Thomson kept using his Faraday tubes, since they were, at the same time structures (thus, explaining discreteness) in the ether (thus, preserving the wave approach). Just to give an example, one image he used to visualize the discrete structure of light supposed that "the ether has disseminated through it discrete lines of electric force and that these are in a state of tension and that light consists of transverse vibrations, Roentgen rays of pulses, travelling along these lines" (Thomson 1907c, p. 421). The energy of the wave would be concentrated in these pulses, thus giving a discrete appearance to the wave-front when traversing a black screen: "the energy of the wave is thus collected into isolated regions, these regions being the portions of the lines of force occupied by the pulses or wave motion" (ibid.). The effect would be, of course, very similar to that given by what he calls "the old emission theory" that spoke of corpuscles of light. The independence from intensity was explained in the following terms: "if we consider light falling on a metal plate, if we increase the distance of the source of light" (ibid.), and considering spherical symmetry from the source, "we shall diminish the number of these different bundles or units falling on a given area of the metal, but we shall not diminish the energy in the individual units" (ibid.).

The latter would explain why the energy of the emitted particle did not depend on the intensity of the incident light. But J.J. had now to explain Erich Ladenburg's finding that the speed of the photo-particles was dependent on the frequency of the light it was irradiated with. He faced this problem by moving a step backwards and incorporating into the picture his theory of the formation of Roentgen rays (which, by "analogy", would be valid for other forms of light): When cathode rays suddenly stopped, they would emit Roentgen rays, the more rapidly moving (thus, more energetic) cathode particles producing thinner pulses (thus, higher frequencies). Thus, the energy present in each unit of light would depend on its frequency, which is what Ladenburg's experiments showed.

Although, 10 years later, Robert Millikan saw this theory as almost equivalent to Einstein's 1905 corpuscular theory of light (Millikan 1917, pp. 221–223), it is clear from Thomson's words that his structured light is perfectly within the bounds of ether physics. It is the physicality of Faraday tubes which allows for this structure of light:

> Thus the structure of the light would be of an exceedingly coarse character, and could perhaps best be pictured by supposing the particles on the old emission theory replaced by isolated transverse disturbances along the lines of force. The greater the frequency of the light the greater is the energy in each unit, so that if it requires a definite amount of energy to liberate a corpuscle from a molecule of a gas, light whose wave length exceeds a particular value, which may depend on the nature of the gas, will be unable to ionize the gas, for then the energy per unit will fall below the value required to ionize the gas (Thomson 1907c, p. 423).

As is well known, the tension between corpuscular and undulatory theories of light would persist until the general acceptance of Einstein's quantum of light and the formulation of a generalized principle of wave-particle duality, both in the mid-1920s (Stuewer 1975). The final acceptance of the corpuscular theory of light came hand in hand with a new, unexpected principle: Louis de Broglie's wave-particle duality. Certainly, Thomson's belief in the reality of the ether, a continuous entity, as the substance constituting electrons, discrete entities, was very far from de Broglie's ideas. Nevertheless, the duality wave-particle reinforced, in a new way, something that had been present in Thomson's work: A belief in pluralistic explanations of matter, each one valid at different epistemic and ontological levels.

4 Conclusion

In his latest book, Hasok Chang (2012) has argued for epistemic pluralism in the sciences. He makes his case on the basis of the history of chemistry in the late eighteenth century up to the early twentieth. From his point of view, the abandonment of phlogiston as a valid—not necessarily true—explanatory tool for many phenomena in chemistry was responsible for a delay in finding a solution to certain problems, problems that were only accounted for after the discovery of the electron. His pluralism is a call against monism in science, against the stance not only that there is one single method in science but also one only set of theories valid at a given time and discipline. His pluralism is related to the points I have tried to make with J.J. Thomson's career in this paper.

The boundaries between physics and chemistry, as well as between any other disciplines, evolved as the result of professionalization and institutionalization of different traditions, methods and questions. Thus, the separation between physics and chemistry into two separate buildings and curricula consolidated both fields as naturally separate. As I have tried to show, this separation was not a necessary one although it was naturalized a posteriori. Cross-, inter-, or transdisciplinarity have been turned into common currency by contemporary policy-makers; but by so doing, they reinforce the naturalness of the separation between "departments of science". The pluralism that Thomson unsuccessfully tried to institutionalize in the Cambridge of his time involved a plurality of methods and objects within a broad department and not the generation of a multiplicity of departments.

This plurality refers also to what could and should be researched. Contemporary scientists smile at the thought of nineteenth-century physicists wasting their time with spiritualism, thought transfer and psychic research. Interestingly, most of these topics are still objects for research in psychology or in neurosciences today. And these researches use mathematical models, electrical apparatuses, etc.

A certain brand of positivism tried to dismiss the aims and methods of the Society of Psychical Research as non-scientific, only to return as legitimate areas for research in the last few decades. Thus, one should be careful to dismiss any potential area of knowledge as irrelevant: The old wish to establish a demarcation between science and non-science proves to be related to the ideologies of particular times.

Finally, the tension between discrete and continuous accounts of matter, between ether and corpuscles, shows that forcing exclusive explanations into the minds of scientists may backfire. Thomson was trying to keep both explanations as valid, only that at different epistemic levels. Quantum theory and the principle of complementarity transformed physics into a more pluralistic enterprise than a generation earlier had allowed for. Certainly, Thomson and de Broglie's approaches were essentially different. But they both coincided in that they tried to keep complementary explanations of the same phenomena rather than sticking to one and exclusive truth.

References

Chang H (2012) Is water H$_2$O? Evidence, realism and pluralism. Springer, Dordrecht

Chayut M (1991) J.J. Thomson: the discovery of the electron and the chemists. Ann Sci 48:527–544

Davis EA, Falconer I (1997) J.J. Thomson and the discovery of the electron. Taylor & Francis, London

Falconer I (1987) Corpuscles, electrons and cathode rays: J.J. Thomson and the "discovery of the electron". Br J Hist Sci 20:241–276

Falconer I (1988) J.J. Thomson's work on positive rays. Hist Stud Phys Sci 18:265–310

Falconer I (1989) J.J. Thomson and "Cavendish physics". In: James F (ed) The development of the laboratory: essays on the place of experiment in industrial civilization. Palgrave Macmillan, London, pp 104–117

Fisch M, Schaffer S (eds) (1991) William Whewell: a composite portrait. Oxford University Press, Oxford

Gauld A (1968) The founders of psychical research. Routledge & Kegan Paul, London

Gooday G (1990) Precision measurement and the genesis of physics teaching. Br J Hist Sci 23:25–51

Harman PM (1982) Energy, force and matter. The conceptual development of nineteenth-century physics. Cambridge University Press, Cambridge

Haynes R (1982) The society for psychical research, 1882–1982. A history. Macdonald, London

Heimann P (1972) The unseen universe. Physics and the philosophy of nature in Victorian Britain. Br J Hist Sci 6:73–79

Kim DW (2002) Leadership and creativity. A history of the Cavendish laboratory, Kluwer Academic, Dordrecht

Kragh H (2002) The vortex atom. A Victorian theory of everything. Centaurus 44:32–114

Maxwell JC (1873) A treatise on electricity and magnetism (2 vols, 3rd edn. 1891). Clarendon Press, Oxford

Maxwell JC (1891) A treatise on electricity and magnetism, 3rd edn. Clarendon Press, Oxford

Millikan RA (1917) The electron: its isolation and measurement and the determination of some of its properties. Chicago University Press, Chicago

Myers G (1989) Nineteenth-century popularizations of thermodynamics and the rhetoric of social prophecy. In: Brantlinger P (ed) Energy and entropy. Science and culture in Victorian Britain. Indiana University Press, Bloomington, pp 303–334

Navarro J (2012) A history of the electron: J.J. and G.P. Thomson. Cambridge University Press, Cambridge

Noakes R (2005) Ethers, religion and politics in late-Victorian physics: beyond the Wynne thesis. Hist Sci 43:415–455

Noakes R (2008) The "world of the infinitely little": connecting physical and psychical realities circa 1900. Studies in the History and Philosophy of Science 39:323–333

Oppenheim J (1985) The other world: spiritualism and psychical research in England, 1850–1914. Cambridge University Press, Cambridge

Schaffer S (1992) Late Victorian metrology and its instrumentation. In: Bud R, Cozzens SE (eds) Invisible connections. Instruments, institutions and science. SPIE, Bellingham, pp 23–56

Sinclair SB (1987) J.J. Thomson and the chemical atom. From ether vortex to atomic decay. Ambix 34:89–116

Smith C (1998) The science of energy. A cultural history of energy physics in Victorian Britain. The Athlone Press, London

Snyder JJ (2006) Reforming philosophy: a Victorian debate on science and society. University of Chicago Press, Chicago

Stewart B, Tait PG (1875) The unseen universe or physical speculations about a future state. Macmillan, London

Stuewer R (1975) The Compton effect: turning point in physics. Science History Publishers, London

Sviedrys R (1976) The rise of physics laboratories in Britain. Hist Stud Phys Sci 7:405–436

Thomson JJ (1881) On the electric and magnetic effects produced by the motion of electrified bodies. Phil Mag 11:229–249

Thomson JJ (1883a) On the motion of vortex rings. Macmillan, London

Thomson JJ (1883b) On a theory of the electric discharge in gases. Phil Mag 15:423–434

Thomson JJ (1894) The connection between chemical combination and the discharge of electricity through gases. Report of the British Association for the Advancement of Science, Oxford 1894. John Murray, London, pp 482–493

Thomson JJ (1895) The relation between the atom and the charge of electricity carried by it. Phil Mag 40:511–544

Thomson JJ (1897) On cathode rays. Phil Mag 44:293–316

Thomson JJ (1899) On the masses of the ions in gases at low pressures. Phil Mag 48:547–567

Thomson JJ (1903) Conduction of electricity through gases. Cambridge University Press, Cambridge

Thomson JJ (1904) Electricity and matter. Scribner, New York

Thomson JJ (1907a) On the light thrown by recent investigations on electricity on the relation between matter and ether. The Adamson Lecture Delivered at the University of Manchester on November 4. Manchester University Press, Manchester

Thomson JJ (1907b) The corpuscular theory of matter. Constable, London

Thomson JJ (1907c) On the ionization of gases by ultra-violet light and on the evidence as to the structure of light afforded by its electrical effects. Proceedings of the Cambridge Philosophical Society 14:417–424

Thomson JJ (1910) Presidential address. Report of the British Association for the Advancement of Science 1909:3–29

Thomson JJ (1914) The forces between atoms and chemical affinity. Phil Mag 27:757–789

Thomson JJ (1925) The structure of light. The Fison Memorial Lecture. Cambridge University Press, Cambridge

Thomson JJ (1936) Recollections and reflections. Bell, London

Warwick A (2003) Masters of theory. The University of Chicago Press, Chicago

Whewell W (1838) On the principles of English university education. John W. Parker, London

Whewell W (1840) The philosophy of the inductive sciences, founded upon their history (2 vols). John W. Parker, London

Yeo R (1993) Defining science: William Whewell, natural knowledge, and public debate in early Victorian Britain. Cambridge University Press, Cambridge

Naturalism and Scientific Hierarchy
An Attempt at Strict Naturalist Normativity

Pedro Sáez Williams

Abstract

The objective of the present chapter is to find a solution that might overcome what seems to be an unassailable philosophical incompatibility between the normative (i.e., epistemological and hierarchical) dimension of scientific practice and the worldview that results from such an endeavor. This incompatibility could also be understood as a conflict between, on the one hand, the a priori assumptions that guide the conduct of scientific practice, legitimize the manner in which such an endeavor is structured (i.e., disciplines and academic hierarchy), and justify its claims, and on the other hand, the metaphysical and/or ontological position that seems to be suggested by the results of empirical research. This is, in so many words, an incompatibility between scientific normativity and naturalism.

Keywords

Social epistemology · Naturalism · Embodied cognition · Normativity · Science

Initially, the term *naturalism* seems to have been used to designate a variety of positions not necessarily continuous with each other. Recent uses of the term denote a call for continuity between science and philosophy. In this regard, ontological scientific naturalism can be adequately described as the commitment to restricting one's ontology (or one's complete inventory of reality) to that which falls under the purview of science, whilst methodological naturalism is a call for

P. Sáez Williams (✉)
Mexico City, Mexico
e-mail: petesaez@me.com

© Springer Fachmedien Wiesbaden GmbH 2016
B.-J. Krings et al. (eds.), *Scientific Knowledge and the Transgression of Boundaries,* Technikzukünfte, Wissenschaft und Gesellschaft / Futures of Technology, Science and Society, DOI 10.1007/978-3-658-14449-4_10

methodological continuity between science and philosophy. The worldview that obtains from the practice of systematic empirical scrutiny, however, is one marked by particular characteristics. That is, scientific results offer a specific outlook on things in which entities and phenomena are embedded in a network of cause and effect relationships. Within this worldview, there seems to be no space for the positing of entities not ontologically contingent on their context.

Methodological naturalism, in turn, at least where it has found success, can be adequately read as the penetration of science and scientific explanations into areas formerly restricted to the purview of alter-scientific disciplines (i.e., religion or philosophy). In practice, this has resulted in the reconceptualization of such previously transcendent items as being causally determined by their context.

For the purposes of the present work, the most important exercise of methodological naturalism has been that of issues pertaining to the organization and justification of science. In this regard, research movements such as the Strong Programme in the Sociology of Scientific Knowledge (SPSSK), the program of science and technology studies (STS), and the embodied approach at the study of cognition (which examines the causal relationship between the body and the mind) have produced scholarship that finds contextual causality and consequence for notions such as scientific rationality, scientific fact, validity, and even logic. By locating the source of scientific legitimacy in the same level of contingency as that which falls under its purview, the naturalization of epistemological concerns has been accused of depriving science (and/or rationality) of its capacity to function as a norm in regards to epistemic issues (Putnam 1982). This, furthermore, results in a challenge to the normative dimension of science as a whole including the organization of scientific practice as a cognitive hierarchy divided in disciplines (Fuller and Collier 2003).

Previous attempts to solve the problem (Fuller 2002; Longino 2002; Kitcher 2001) have had to forgo a strict rendering of naturalism and settle for a solution that allows the meta- or super-natural in either ontological or epistemological terms.

An account of normativity that may provide science (or any other organized endeavor) with ideal guidance or prescription without falling into any form of natural transcendentalism would have to adhere to the following:

1. It would have to be sourced from scientific evidence,
2. also, maintain all of its ontological posting within the bounds of that which falls under the purview of scientific explanations (namely entities and phenomena contingent on causal-consequential relations to their spatiotemporal context),

3. abstain from assuming any source of normative import for its determinations that transcends the possibility of revision, correction, scrutiny or change within the framework it creates, and, finally,

4. be capable of providing normative import beyond its context (i.e., normative import for third parties).

The main purpose of the present work is to provide the means for such an account.

1 Strict Naturalism: Ontological, Methodological and Epistemological

A particular worldview results from abstaining from positing the presence or existence of phenomena and entities prior to the systematic scrutiny of reality or nature. In other words, a particular worldview results from maintaining one's ontological inventory within the boundaries of that which results from the practice of science. Entities and phenomena, in this worldview, are hypothesized as explainable. Explanations (at least not circular ones) do not account for entities in virtue of themselves (i.e., A because of A); they, rather, require accounting for the nature of entities or phenomena, in virtue of their relationship to other entities or phenomena. In this regard, the worldview that obtains from the accounts of science is one of causality-consequence relationships or, in other words, of ontological inter-dependence. For example, a first cause (i.e., the creator) cannot be explained but in virtue of itself: It is its own cause and its own explanation. A scientific account of tides, on the other hand, explains these by reference to the gravitational forces exerted by the moon and the sun in combination of the Earth's rotation. Said forces, in turn, are also accounted for in regards to other causes and so on, and so forth.

This position—the commitment to restricting one's ontological positing or one's ontology (understood as one's complete inventory of reality) to that which falls within the purview of science—has been called ontological naturalism (Rudder Baker 2013, p. 5; De Caro and Macarthur 2004, p. 3; Kornblith 1994, p. 40; Sellars 1963). Historically speaking, the term can be traced back to Spinoza's (1996 [1677]) pantheism, or the position that God and the physical Universe ("Nature") are a single unified substance (Fuller 2007b, p. 106). Modern uses of naturalism conserve this understanding of the term in relation to physical or material immanence. In this sense, the meta- or the super-natural is understood as that which transcends the physical domain. This would obviously include posited entities such as ghosts, goblins and gods (Ritchie 2008, pp. 2–5), but also abstract entities such as numbers (in a universalist sense), other mathematical abstractions, Platonic forms, etc. This means, in so many words, that anything that is posited is located "outside or 'extra to' space and time" (Armstrong 2010; Melnyk 2003, p. 9).

Both of these understandings, however, are continuous with each other. For example, entities and phenomena whose ontological attributes are contingent upon their relation with other aspects of their domain are not, by definition, transcendent of said domain. Likewise, that which is transcendent or ontologically independent cannot by definition be ontologically contingent on anything else. It could be said then that ontological naturalism is, in overall terms, a commitment to ontological monism or horizontality in the sense that it rejects the positing of entities and phenomena that somehow occupy a higher status in regards to the ontological circumstances of the rest, either on account of being a first cause or otherwise able to escape the network of ontological contingency observed in "nature".

A related use of the term "naturalism", one that has come to be emphasized in contemporary academia, is as an appeal for methodological continuity between science and philosophy at the expense of the latter (De Caro and Macarthur 2004, pp. 3–9). Naturalism can thus be understood as a call to subject phenomena and entities that had been previously accounted for by purely philosophical or otherwise alter-scientific means (such as the mind, political rights, or the notion of knowledge) to the purview of systematic empirical scrutiny and, thereby, within the reach of causal explanation. Methodological naturalism in this sense can be fortunately understood as a historical tendency in which natural philosophy and later science has gradually displaced meta-natural or alter-scientific explanations for a wide range of phenomena.

Much of what was displaced, especially in early attempts, would count as what is commonly labelled as religion or religious. The most notable example of this is the displacement of transcendental theism by theories that suggest that life is governed by natural causality (i.e., Charles Darwin's *The Origin of Species*). The most interesting displacements of transcendental accounts are not, however, related to religion. A case in point is the Marxist explanation of liberal ideology, particularly that of the notion of individual liberties—which had until then been understood as universal and self-evident—, as an integral element of the legal and political superstructure that results from (and is therefore *causally* related to) a social arrangement based on capitalist modes of production.

As can be deduced from the above description and inferred from the given examples, once something that had previously been excluded from the purview of science is then placed in its sphere, it acquires, in consequence, the characteristics of nature, namely a place within the realm of causal-consequential relations and ontological contingency.

That which falls under the purview of epistemology is no exception. That is, when matters pertaining to epistemic value and/or knowledge are the subject of systematic empirical scrutiny, they are found to be as contextually contingent and as causally determined as any other entity or phenomena that occupies a place within the realm of nature. One of the most notable examples of the above is the sociology (and overall social study) of (scientific) knowledge.

Scientific knowledge, traditionally understood as being valid regardless of context, was initially spared the scrutiny of the sociological endeavor. In the words of Camic and Gross (2001), the "old sociology of ideas" (i.e., Mannheim 1929; Parsons and Platt 1973; Merton 1970) "assumed unproblematic distinction between the content of ideas", that is their "internal substance" and any and all "external factors". These authors contend that within the methodological framework of the old sociology of ideas, the substance of scientific thought was assumed to be "in varying degrees the realm of an asocial, scientific rationality about which sociology could have little to say" (Camic and Gross 2001, pp. 238–239). It is in this sense that Karl Mannheim, for example, contended sociology was only suitable for the explanation of certain forms of knowledge, such as religious, moral or social knowledge, but not for the "exact sciences" (Mannheim 1929, p. 43).

In this sense, it is of interest to note, that the methodological exclusion of these matters from sociological (or otherwise social-scientific) scrutiny translates into the assumption that the content of scientific knowledge requires no contextual explanation and is, thereby, causally independent of all social phenomena or context. For these studies, the cause or source of scientific knowledge (if they assume that the production or possession of such knowledge is attributed to anything at all) must then be something that is also independent of context. Examples could include "method", "rationality" or other means of acquiring truth or validity transcending a situation.

Once this a priori methodological constraint is no longer operative and, thereby, scientific knowledge and matters pertaining to its production and justification are placed under the purview of systematic empirical scrutiny, phenomena such as epistemic value and knowledge show characteristics at odds with those described above. Early examples of the above are found in the work of Ludwik Fleck, who as a result of empirical research on scientific practices claimed that truth and perception were relative to what he termed *thought collectives* or *Denkkollektive* (Fleck 1979 [1935], p. 100). More recent work in the same vein is found under what is commonly grouped under the heading of the Strong Programme in the Sociology of Scientific Knowledge (see Barnes et al. 1996; Barnes and Bloor

1982). The best received and also most controversial research movement continuous with a methodological commitment in line with those described above, however, is that which is commonly grouped under the designation science and technology studies (STS). STS proceeds under a methodological naïveté that constitutes an attempt to forgo all preconceptions, thereby eliminating all a priori assumptions that could lead to certain phenomena being excluded from systematic empirical treatment. One of the most important claims that result from the bulk of ethnographic work classified under STS (e.g., Latour and Woolgar 1986; Latour 1988; Knorr-Cetina 1999) is that scientific facts are created by contingent action. In their words, "the assumed 'out-there-ness' [of 'scientific reality'] is a consequence of scientific action rather than its cause" (Latour and Woolgar 1986, p. 182).

The naturalization of epistemological matters, however, is not the exclusive result of the empirical scrutiny of phenomena pertaining to knowledge production and justification. Arguments leading to a call for epistemology to be reduced to a chapter of empirical science, and thereby to causal-consequential explanation, have also emanated from philosophy itself. Such is the case of W.V.O. (Willard Van Orman) Quine's (1969b) *Epistemology Naturalized*. Quine calls for the naturalization of the epistemological endeavor because he finds no philosophical warrant whatsoever for advancing (or elaborating) principles whose epistemological firmness is such that they transcend the possibility of being changed upon revision.[1] If all knowledge is subject to the possibility of future correction, there are

---

[1] Also called the rejection of the synthetic-analytic distinction, Quine sustains this position on a variety of negative philosophical arguments (see Quine 1969b, 1975); all of these are sophisticated versions of historically powerful arguments against the possibility of epistemological (or otherwise metaphysical) certainty such as: (1) *the inscrutability of alter experience (or perspective)* which questions the epistemic certainty required to stipulate (normative or otherwise) accounts that transcend its author, in virtue of the impossibility to determine if the very same arguments would still hold the assumed validity from a different perspective (the perspective of your listener or receiver perhaps); and (2) *the inscrutability of the future*, or the realization that future knowledge is by definition unknown and therefore unforeseeable, therefore something that is "not known" may (from a realistic perspective) shed light into past errors once it is "known", making it therefore impossible to ascertain what elements of our theory are true and which are erroneous. Either of these two positions provides for an extremely simple but illustrative rendering of Quine's (1951) rejection of the analytic-synthetic distinction. This means, either position points to the realization that all knowledge is subject to revision (either by future experience or by communication with an agent of alter experience). And so, even if one acknowledges the a priori validity of some sort of rules or norms that dictate one's reasoning (one might call *this form* "logic"), one must thereby assume validity to all that results from the computations of such rules, including the challenges that question a certain universally stable and non-temporal validity of the logical laws that one is acquainted with (such as the two principles stated above). Not

then no grounds for principles being prior, firmer or epistemologically higher than that which results from scientific scrutiny. This implies a rejection of the possibility of the establishment of a meta-theoretical framework for the prescription of knowledge production and/or a standpoint for the judgement of epistemic validity.

The rejection of a meta-theory or, in Quine's (1981) words, a "first philosophy" translates into a commitment of "epistemic horizontality". In traditionalist epistemological accounts (i.e., epistemology/methodology or philosophy of science), the value of scientific theories is *contingent* on their adherence to the principles of a meta-theory upon which all other theories are to be judged. Even if such principles were (within such a philosophy) not considered to constitute "transcendent truths", they would, nonetheless transcend the possibility of scrutiny (and thereby revision or correction) by the framework they create, precisely by being the foundational basis for such a framework (in a way not dissimilar to the manner in which a "first cause" transcends the network of causality of its own creation).

By eliminating a prior or higher framework for the determination of a higher value or reference, epistemic horizontality commits itself to a relativist understanding of reference and (epistemic) value: The meta-theoretical perspective is precisely the rejection of a meta-theoretical perspective. Both of these attributes are necessarily contingent on a particular theoretical perspective that cannot be valued as being higher or lower than any other perspective. In Quine's words, "reference is nonsense except relative to a coordinate system" (Quine 1969a, p. 48).

If ontological naturalism can be understood as a commitment to ontological horizontality that results from maintaining one's ontological positing to that which falls under the purview of systematized empirical scrutiny, epistemological naturalism (at least in Quine's strict rendition) can be understood as the commitment to epistemological horizontality that results from the realization that no form of knowledge escapes the possibility of revision, correction, and change.

Whilst the ontologically naturalist perspective of knowledge and epistemic value/justification (the view that results from STS) finds these to be contingent on a specific ontological context (i.e., social or cultural context), the philosophical account of naturalist epistemology (Quine) finds these very same contingent on a determined theory, theoretical perspective, or coordinate system. In this sense, one

---

(Footnote 1 continued)
doing so would imply an arbitrary choosing of logical validity contingent to epistemological convenience. Because all knowledge, including logical form or any other assumed parameter of a priori validity (such as standards of rationality), is subject to revision, there is then no philosophically important distinction between synthetic or analytic knowledge.

may speak of a certain degree of continuity between these approaches, whose extent (the continuity's), however, is (from the above) difficult to grasp.

Regarding that which they reject, the continuity between the practical and the philosophical dimensions of naturalized epistemology is, nevertheless, easier to appreciate. First philosophies (i.e., methods and/or accounts of rationality, etc.) seem to be exactly that which the social studies committed to the internal-external distinction assume is the cause of scientific knowledge and/or its source of the epistemic validity. In this sense, once scientific knowledge is naturalized, the firmer epistemological ground of first philosophies that provides their assumed epistemological firmness turns out to have a direct ontological equivalent in what Nagel (1986) calls a "view from nowhere".

2   Naturalism and Normativity

Metaphysical horizontality (both ontological and epistemological), however, seems to find itself at odds with all attempts at normativity. As discussed above, Quine's rejection of a firmer standpoint from whence to judge epistemic value leads to a rejection of the epistemological endeavor understood as a normative project (Quine 1969b, pp. 82–84). The scientific process, for Quine, is an inherently circular enterprise in which one proceeds from "the standpoint of a theory building process" towards the standpoint of "a theory being built" (Quine 1960, p. 22). All assumed validity is relative to the position of the presently accepted theory. Scientific justification, a meta-theoretical endeavor, is beyond possibility and thereby considered to be, from a Quine (1969b) perspective, futile and unnecessary; naturalist epistemology (understood thusly) is, thereby, just an attempt at understanding (Hylton 2007, p. 83; Quine 1969b, p. 83).

Ontologically speaking, furthermore, norms have historically also had hard time locating themselves within nature (Hume 1888 [1970]). That is, within an ontological inventory constituted entirely by entities and phenomena located within the domain of causality, empirical scrutiny and change, there seems to be then "no room for normative facts—or at least be so insofar as they cannot be reduced to the kinds of objective, causal facts with which the natural science deals" (De Caro and Macarthur 2010, p. 1).

In general terms, the problem between naturalism and normativity seems to be that the overall metaphysical (ontological or epistemological) firmness that a first philosophy is required to possess in order to have any normative import in regards to that which is purported to fall under its sway seems, further, to be required by *all* attempts at normativity: In order to achieve normative import, that which functions as a norm must be firmer than that which is to be normed. Norms, it seems, must not be subject to the same contingencies, otherwise, their value determinations and/or prescriptions would be as transient as the characteristics of that which is to be judged, prescribed or valued. In so many words, normativity (understood thusly) would seem to require commitments to metaphysical hierarchy that are not continuous with naturalism's commitments to metaphysical horizontality.

Many have, on similar grounds (see De Caro and Macarthur 2010; Putnam 1982), contended that a strict rendering of scientific naturalism is incompatible with normativity. In order to maintain its commitments to metaphysical horizontality in either ontological or epistemological terms, strict naturalism would seem to be restrained to fallibilist or hypothetical causal realism, that is to the positing of causal-consequential explanations without being able to judge in regards to their epistemic value.

Science, nevertheless, seems to be a profoundly normative endeavor, which acts in ways at odds with those described above. First, science is a methodological and therefore prescriptive endeavor (Ritchie 2008, p. 74). As explained above, this would not be at odds with naturalism (specifically its epistemological component) if science abstained from judging the epistemic value of that which is obtained from the adherence to its methods. This, however, hardly seems to be the case.

It is hardly controversial to sustain that science, understood as the formal systematic effort at the production of knowledge, proceeds under the assumption that the adherence to such methodological prescriptions does lead to the production of epistemological value, and that these prescriptions provide an objective framework for epistemic discrimination. Science, for example, distinguishes itself and its claims from the claims and practices of other knowledge producing ventures that are deemed, by its own (methodological) standards, as pseudo- or un-scientific (examples might include homeopathy and astrology).

Science, also, does not seem to pronounce these deliberations in the first person or allude to their contextual validity. Science, rather, speaks in universal terms: It treats its pronouncements as facts, which, furthermore, are received by wider society (both tacitly and formally) as such, leading to a normative relationship between science and society at large (Fuller 2002, p. 177).

Finally, on account of the above, scientific (epistemic) judgements have practical, economic and political consequences. Expert or scientific knowledge, for example, plays an important part in the diagnosis and determination of possible courses of action in regards to both individual problems and social issues (Rouse 1987, p. 227). In formal terms, the epistemic determinations of the scientific establishment constitute the foremost authority concerning the standardization of the education curriculum and health provision (Fuller 2000, p. 8; 2002, p. 177).

Also, scientific epistemic determinations play an important role regarding the determination of the formal legal verdict on what counts as "objective", "secular", "valid" or "neutral" statements and claims to knowledge.[2]

Regarding economic matters, scientific valuation also has consequences that extend well beyond the bounds of the scientific establishment. The scientific enterprise requires vast amounts of resources to conduct its activities. In this regard, and in lieu of their recognition of epistemic (cognitive) authority, scientists as a class are able to procure vast amounts of economic capital and manpower both to conduct their professional activity and to extend the products of their work into the domain of laymen (Rouse 1987, pp. 212–226).

From the above, a relationship between science and the rest of society, or between formal experts and laymen, can be inferred. It seems to be the case that the epistemic judgments of science or scientists (hereafter [a]) have import for the capacity of action of wider society or laypersons (hereafter [b]). However, it seems that the determinations of (a) are assumed to be valid before or independently of the participation or position (situation/location/context) of (b). The source (i.e., method) of the authority that (a) is capable to exert in regards to (b) must, therefore, be assumed to transcend (b)'s situation/location/context (Fuller 2002, pp. 175–189; 2007a, pp. 43–52).

This situation also implies a tacit and formal socio-political hierarchy in regards to cognitive or epistemic matters. Since it is implied that the "expert" (a) is—in objective or realist terms—more knowledgeable than the layman (b), then (a) has a higher cognitive or epistemic ranking than (b), also in objective or realist terms. Given this understanding of cognitive authority, the ideal is then to formally reproduce this assumed factual cognitive hierarchy. In practice, it means that members of the cognitive hierarchy are selected from the top down. In other words, those (a) that, as a result of their future formal relation to science, will acquire the capacity to act in ways that will affect others (b) are not selected by the bulk of those possibly affected (b), but rather by those who already have a high rank in the cognitive hierarchy (a+). The choosing of future experts and scientists is done by, and according to standards developed by, firmly established experts and scientists "through examination and publication policies that proceed with little external scrutiny" (Fuller 2000, p. 8).

---

[2]This is evidenced by a plethora of examples, of which the most significant are: (1) The admissibility of expert testimony for both judicial and administrative matters in most secular states is commonly regulated by procedural legislation that defers to scientific standards (i.e., Rule of Evidence 702 in the U.S.A., see Bernstein and Jackson 2004).

An expert (a) in one area is a layman (b) in another, leading to the guiding principle for setting disciplinary boundaries. Academic disciplines, in this sense, can be understood as the horizontal dimension of the cognitive hierarchy (Fuller 1997, pp. 49–59; 2002, pp. 191–195).

In short, the internal organization of science as well as its formal relation to the rest of society presupposes an epistemic, or cognitive, hierarchy in which certain individuals ([a], experts) have a higher standing than others ([b], laymen) in virtue of a source of authority that is independent of both. To refer again to Nagel's (1986) expression, "view from nowhere": A perspective ontologically (and epistemologically) located beyond that which it observes (nature).

If the above were a fortunate approximation of the current organization of the scientific enterprise, then the epistemological assumptions that (supposedly) norm the scientific establishment (and justify science's normative dimension) would seem to be at odds with the general worldview that obtains from its practice. This purely philosophical contradiction is, however, the least pressing of the possible negative consequences that might result from a hierarchical organization legitimized by assumptions of political transcendentalism. In this sense, the movement commonly grouped under the heading naturalist social epistemology (or NSE; see Fuller 2002, 2007a, 2009; Fuller and Collier 2003) claims that such a situation may result in important consequences for the capacity of wider society to produce knowledge (understood as the capacity to procure new capacities of description and action) as well as other significant consequences of political nature.

The NSE, founded on reasons such as those mentioned above, takes seriously the possibility that a source of authority that is un-revisable and non-contingent to the causal forces of its spatial-temporal context and is therefore un-situated in nature (i.e., a "view from nowhere") is unwarranted. If this is the case, and science's validity is different to that which legitimizes its cognitive hierarchy, then in a situation such as that described above the primary incentive of science's highest ranking members (a+) is to maintain or further the hierarchical structure of the establishment, in spite of the consequences this may have in regards to the production of knowledge. This may translate into the following consequences:

- The scientific hierarchy is able to set the evaluation criteria of its own activity, that is, the appropriateness of methods. As the benchmarks of progress, or improvement, are altered to benefit the current orthodoxy, the history of standards is altered as well to make it appear that the current situation would also seem fortunate from the perspective of previous times (Fuller 2000, p. 10; 2002, pp. xx–xxi; 2007a, p. 25).

- Regarding new players, that is, new theories or research approaches that might seriously challenge the current orthodoxy and therefore the hierarchical arrangement of the discipline in question, the lack of external scrutiny allows the established structure to decide how they will incorporate, or import, theories into their research. In this sense, it is then in their best interests to "capitalize on their initial conceptions, and ignore—not test—the theories that challenge" them (Fuller and Collier 2003, p. 32). Furthermore, as scientific research begins to increase in terms of scale, that is, in terms of "size, complexity, hierarchical level and [required] material investment", the incentives to allow challenges are reduced dramatically (Fuller 2000, p. 29).
- If, however, a significant challenge is being posed from *within* (scientific hierarchy) or a specific scientific movement manages to acquire the social, political and economic capital required to compete against the established status, the competitor will most probably not disappear, but rather a *new discipline* will find its place amongst the others. Two disciplines may then have "orthogonal" objects of study (such as cognitive psychology and neurophysiology). This is equivalent to a spatial (as opposed to a temporal) paradigm shift in Kuhnian terms Kuhn (1970). In this sense, "disciplinary boundaries many [sic] be seen as fault lines that conceal future scientific revolutions" (Fuller 2002, p. 195). The maximizing of the pool of funds available would not necessarily translate into an increased capacity of knowledge production but rather "enable all to continue on their current trajectories as they see fit" (Fuller and Collier 2003, p. 43).

All true scientific progress can be understood as a transgression of boundaries, since it is precisely the scientific hierarchy in its horizontal dimension that serves as the most pressing social epistemological problem. This does not imply that, under the organizational assumptions at hand, scientific research will not lead to *any* results that might translate into practical benefits. It does mean, however, that the incentives are not geared to that purpose, but rather to the maintenance of the status quo, the accumulation of resources, and the increased cognitive separation between expert and layman. As this takes place, the latter begins to rely increasingly on the opinions of experts, to the point that a greater number of issues are increasingly removed from the arena of public debate and placed under the testimony of scientists (Fuller and Collier 2003, p. 11), which leads to the main political problem that arises from this situation, namely that

> [...] carried to its logical extreme, cognitive authoritarianism of this sort would claim that the only decision that the public is entitled to make is to fund more social scientific research to determine the identity of the 'public interest' from the many misleading things that people say and do (Fuller and Collier 2003, p. 287).

In lieu of the above, the question arises then as to how science is to be organized. The philosophical and practical consequences that obtain from taking epistemological naturalism seriously have been addressed in various ways. Some such as Quine (1969b), STS and SPSSK retreat from the (categorical) normative endeavor. Most (e.g., Putnam 2010; Longino 2002; Kitcher 2001) deny the causal-contextual nature of knowledge in one way or another, acknowledge the results that suggest it in other ways, and find a solution in between.

Those who attempt to maintain themselves within the bounds of strict *epistemological* naturalism, namely the NSE, appeal to the elaboration of a republican constitutional regime for science, in which the currently assumed transcendental authorities (the source) are replaced by socially sanctioned counterparts that mediate between knowledge producers and those with an economic, political and practical interest in the production of knowledge (namely, society at large). The general spirit behind such a suggestion is to "recast disciplinary boundaries as artificial barriers to the transaction of knowledge claims" (Fuller and Collier 2003, p. 31), and to place researchers "in direct competition with one another where they previously were not", in order to force them to account for their findings "not only to their own discipline's practitioners, but also to the practitioners of other disciplines and maybe even the public" (Fuller and Collier 2003, p. 23).

The problem with Fuller's proposal is that it seems to be afflicted with the same issue that his negative account attempts to address. That is, Fuller's strong naturalist commitments are hardly compatible with his transcendental commitments to constitutionalism and his own version (termed "civic republicanism" [Fuller 2000]) of (epistemic) social democracy. In other words, Fuller trades transcendental epistemological commitments for an a priori valorization of specific political theories or ideologies.

The above-mentioned attempts to provide science with normative guidance—in spite of or by taking into account the results of the naturalization of epistemological matters—can be reduced to renditions of what has been called *liberal naturalism* (De Caro and Macarthur 2010), namely *a commitment to ontological and epistemological naturalism in regards to all that does not interfere with what is necessary for one's attempt at normativity, and an allowance of transcendentalism in regards to that which does.*

Provision of a means to organize and prescribe the conduct of science whilst maintaining oneself within the bounds of a *strict rendering of naturalism*—such as that described in the preceding sections—would imply providing a normative account that:

i. was sourced from scientific evidence,

ii. maintained all of its ontological posting within the bounds of that which falls under the purview of scientific explanations (namely entities and phenomena contingent on causal-consequential relations to their spatiotemporal context),

iii. does not assume any source of normative import for its determinations that transcends the possibility of revision, correction, scrutiny, or change within the framework it creates, and

iv. is capable of providing normative import beyond its context (i.e., normative import for third parties).

What follows is precisely an attempt to provide such an account.

3   Horizontal Normativity: An Attempt at Strict Naturalist Social Epistemology

Considering the seemingly intuitive relationship between knowledge and cognition, the cognitive sciences would seem to offer the most fertile ground to find a source for a naturalist attempt at scientific normativity. In this regard, it is interesting to note the presence of a similar situation in the cognitive sciences as that observed in the sociology of knowledge: Most of the traditional approaches to the study of the mind have a priori methodological commitments that translate into the attribution of transcendental characteristics to cognitive phenomena. These are commitments that lead to phenomena being posited as un-situated or non-contingent on the spatiotemporal characteristics of its context.

For example, the elements that have been collectively grouped under "cognitive computationalism" (Thompson 2007, p. 10) or "cognitive realism" (Varela et al. 1991, p. 147), represented by the research programs of cognitivism (Pylyshyn 1984; Fodor 1987, 1985) and connectionism (Hofstadter 1995; Dennett 1978), commit themselves to an a priori understanding of cognition as the processing of symbols or representational stand-ins for a stable and prior reality whose semantic content is causally related to this exterior.

Initially, this position would seem to provide for a naturalist account of realism and therefore a (scientific) foundation for (categorical) scientific realism (Goldman 1993). It is hard to argue, however, in favor of the scientific credentials of such a position when the quality that defines it and would seem to provide empirical support for a scientific account of scientific validity has been determined a priori as methodological commitments that lead to a transcendentalist understanding of the mind (not dissimilar to those proposed by Cartesian dualism). For example, such a position assumes both (1) universalism in the sense that concepts or tokens are caused by a reality that is both previous to and independent of the cognitive agent's spatiotemporal situation (e.g., Fodor 1987; Pylyshyn 1984, p. 50); and (2) the lack of a spatiotemporal situation and determination of the rules that determine the computations at hand. Even cognition is thus assumed to be causally related to external reality, to concepts themselves and to the rules that determine computations held ontologically irrespective of the situation or spatial-temporal location. In this sense, Fodor (1985) acknowledges that "the *only* respect in which [realist] contemporary cognitive science represents a major advance over the RTM [representational theories of mind] [of] its eighteenth- and nineteenth century predecessors" is its use of the "computer metaphor" (Fodor 1985, p. 93; also see Chemero 2009, pp. 3–17).

The theoretical work grouped under the heading "radical embodied cognition" or "embodied dynamicism" (Gallaguer and Zahavi 2008; Gallaguer 2005; Thompson and Varela 2001; Gelder 1995; Varela et al. 1991), on the other hand, abstains from ontological claims that might interfere with the interpretation of empirical evidence. The evidence that specifically touches on the matter, furthermore, seems to locate all cognitive phenomena within the bounds of space-time and ontological contingency. An unsurprising situation considering what seems to be required for scientific explanations (see Sect. 1).

For example, since the 1800s, scientific work on color vision (Young 1802; von Helmholtz 1970 [1866]) has hypothesized that the phenomenal structure of human color perception results from a limited number of perceptual parameters. Empirical work (Brown and Wald 1964) later confirmed that phenomenal color structure was determined by anatomic characteristics, specifically the number of visual pigments found in the retina, which in the case of humans are commonly three: Short-wave, or blue (capable of absorbing maximally at wavelengths of 430–450 nm of the electromagnetic spectrum), middle-wave, or green (540–550 nm), and long-wave, or red (555–575 nm). The most accepted account of animal color vision, the "dual-process theory of colour" (Hurvich and Jameson 1957), draws on findings such as the above and hypothesizes that color phenomena are the result of the structural arrangement of hues in opposing pairs (i.e., black/white, red/green) based on the number of parameters (visual pigments) found in an organism. Color then varies significantly and is contingent, amongst species and individual organisms, on the number of visual pigments observed and their spectral sensitivity. For example, birds such as ducks and pigeons are tetrachromats (possibly pentachromats) (Bowmaker 1980, p. 196; Goldsmith 1990).

Work concerning the ontologically contingent (embodied, sensory-motor grounded, etc.) nature of perception is still dominated by research and arguments revolving around color experience (Wessel 2006, p. 93). There is, however, some incipient empirical (or semi-empirical) work concerned with the situatedness of auditory experience (Wessel 2006; Lochhead 1995), and there has been important and continuous work related to the sense of smell (Skarda and Freeman 1987). The overall trend of these studies is that, in general terms, the phenomenal structure (the structure of perception) is contingent on the specific characteristics of the body (and, therefore, sensory-motor capacity). In other words, *experience is ontologically contingent on anatomy*. Because sensory-motor capacity is in constant change (either in genetic [evolutionary] or ontogenetic [even cognitive] terms), experience is also contingent on time or rather on the temporal situation. Of most importance, however, is the realization that the phenomenal structure of two agents—depending on their specific characteristics—may be incommensurable. For example, in terms of color dimensionality, agents seem to have non-comparable experiences: Two different color spaces cannot be mapped or translated from one to the other without remainder (Matthen 2005, pp. 184–185; Thompson 1995, p. 151).

So far, this line of work has been concerned with what is called on-line cognition, that is, cognitive phenomena constituted by a direct relation to the environment (i.e., sensorial experience or perception). There is substantial work, however, that suggests that a similar situation is present in the case of off-line cognition (i.e., abstraction, imagination, representation). In this regard, evidence in fields such as linguistics and psycholinguistics (e.g., Gibbs 2003; Lakoff and Johnson 1999) and neurophysiology (e.g., Kiefer and Barsalou 2013; Barsalou 2008) suggest that cognitive phenomena such as language, imagination and overall abstraction ("representation") are also grounded (Barsalou 1999), embodied (Lakoff and Johnson 1999) or otherwise contingent on the organisms' perceptual and motor mechanisms.

Some of the most stimulating research in this regard stems from studies in the field of neuropsychology that suggest a strong correlation between abstract reasoning and the neural mechanism used in perception and movement (see Kiefer and Barsalow 2013). For example, studies have found that sound-related concepts quickly (150 ms after word onset) activate an auditory-associated cortex (Kiefer et al. 2008). A related example comes from Pulvermüller (2005), who found a similar correlation between action-related concepts and motor areas.

Findings such as these have prompted theorists to hypothesize that all abstract (representational or symbolic cognition) is grounded or constituted by sensori-motor simulation: The areas of the brain that are used for perception and action are the same areas used for abstract cognition (Lakoff and Johnson 1999, p. 20; Barsalou 1999, pp. 582–583), including mathematical reasoning (Lakoff and Nuñez 2000).

It is then interesting to note that these studies operate as an individualist counterpart for the STS: Whilst STS suggests that scientific knowledge is determined by the causal and spatiotemporal situation of a specific cultural or social context, studies such as the above suggest that cognitive phenomena—in general terms and including scientific knowledge—are also situated within the domain of causality, time/change and naturalist explanation, in this occasion, however, within the bounds of organism anatomy. Furthermore, by suggesting that all cognitive phenomena are situated and that cognitive phenomena vary in ways that are even incommensurable, they also challenge the possibility of an un-situated higher ground on which to norm true from false, the valid from the invalid, or the real from unreal.

Embodied dynamicism can be understood as the ontologically uncommitted theoretical approach that stems from the results of the afore-mentioned studies. The term, as used by Thompson (2007), highlights the theoretical continuity between two programs of different origins: The dynamical systems approach to cognition (e.g., Thelen and Smith 1994; Gelder 1995) and radical embodied cognition (Chemero 2009; Thompson 2007; Varela et al. 1991). A wide variety of work differing in regards to specifics and main focus of study can be grouped under these headings. In general terms, however, they are all, in one way or another and to different degrees, committed to a statement such as the following:

> Cognition, in general terms (this is including perceptual "experience" and "abstract thought" or its equivalent), is both *relative* to and *constrained* by the ultimate spatiotemporal situation. The spatial dimension is represented by the physical, functional and organizational characteristics of the body. And the temporal dimension is represented by the body's genetic, ontogenetic and social history.

In order to explain how a normative account continuous with strict naturalism may be obtained from such a position, it is necessary to explain its two main components. The first is that cognition is *relative* to a natural location (ontologically/causally contingent on anatomy and/or sensory-motor capacity). This is equivalent to a rejection of the "view from nowhere" and the impossibility of a higher or firmer (meta-cognitive) ground to judge the fortune of cognitive phenomena.

The second position is that cognition is *constrained* by natural location. This is a related but different claim from the above. It implies that cognition, even though *relative, is not arbitrary*. For example, it may be the case that an object that I (being a trichromatic being) perceive as (what I have been culturally instructed to label as) blue is perceived by another (possibly pentachromatic) agent as colored by three different hues which he has been culturally instructed to label as X, Y and Z (differences in chromatic dimensionality imply the perception of more "hues"). The fact that I perceive blue and my alien friend perceives X, Y and Z, does not imply that we have a choice in the matter. Our perception is then both *relative and constrained* by the ultimate spatiotemporal characteristics of our phenomenal cognitive structure, specifically the relationship of anatomy and our environment. In this particular example, constraint means that our *cognitive situation transcends the (direct) volition of cognitive agents.*

Consider, in this regard, Thompson's (2007) elaboration on the relationship between, on one hand, the specific sensory motor capacities of a bacteria (its metabolic needs) and, on the other hand, a specific aspect of the bacteria's environment, sucrose.

> When swimming in the presence of a sucrose gradient [bacteria] will tumble about until they hit upon orientation that increases their exposure to sucrose (Thompson 2007, p. 74).

The bacterium acts in such a way because its requirement of sustenance results in a specific categorization of the environment. Certain aspects of it, then, have *functional significance* to others, and in regards to its metabolic needs, certain aspects of the environment acquire *value* (metabolic value). Living structures, according to the author,

> [...] are thus ontologically emergent with respect to mere physical structures. They constitute a new order of nature that is qualitatively distinct from the merely physical order. [...] [A characteristic of] the living order is that the relation between organism and environment is meaningful and normative (Thompson 2007, p. 74).

Value (functional or otherwise), from such a naturalistic perspective, is a result of the necessary categorization, differentiation, and discrimination of the environment. Life is equivalent to evaluation and significance and, therefore, *life is normative.*

Whereas for the biological requirements of the bacteria the whole evaluative domain may be encompassed in functional terms (i.e., by the duality of attractiveness and repulsiveness), the evaluative domains of more sophisticated organisms may in turn also be laden with greater sophistication. And so, for example, a meta-zoan will categorize his environment in regards to various different sensorial standards. The above-described work on color is a prime example of this phenomenon.

As has been the matter of the present work, in the case of humans, we find that one of the most (if not the most) important aspects of our evaluative domain is that which concerns epistemic matters. Like any other biological valuation, however, *epistemic valuation is also relative and constrained to a situation (not arbitrary)*. Whilst this might initially seem a bold claim, it is actually equivalent to a relatively uncontroversial position within analytic philosophy, namely doxastic involuntarism or the notion that "[cognitive agents] lack [...] direct voluntary control over [their] beliefs" (Alston 1988, p. 260).

In specific terms, this implies that one cannot alter one's doxastic state (cognitive state in regards to belief) without engaging in action, and even so one cannot control the effects that such action may have on such as state. As an example of this situation, Alston (1996) requests the reader to attempt to voluntarily believe that the USA is still a colony of Great Britain. In my case at least, I am unable to do so. I am unable to believe at will.

If one has no direct control over the one's relationship between doubt and certainty (what I call one's *limit of certainty*) regarding any particular matter, this implies then that (for practical purposes) one has (ultimately) no direct control over the degree of cognitive authority that one awards to cognitive (intellectual) positions, experience, and/or communication of positions and statements. In other words, one does not have direct control over one's epistemic judgements. It is precisely this experienced constraint which allows for a normative account that is both compatible with naturalism and may provide the ground for a normative framework for science.

The first step to reach such a goal lies in the notion of *situational necessity*: A modal qualification that does not claim authority from any metaphysically transcendent source.

In order to arrive at the manner in which situational necessity operates, a review of where we currently stand is necessary. So far we have found two insurmountable positions:

(1) *The (ontological, epistemological or phenomenological) relativity of experience, knowledge and cognition or the inscrutability of the impossibly experienced (IIE)*: For both philosophical and empirical reasons it has been concluded that a (normative, epistemological or empirical) perspective that transcends causality/possibility of explanation and time/change is unwarranted (by any presently known standard). This means that there is no warrant to claim for an independent (non-contingent) standpoint on which to judge reality; no standpoint to determine real from unreal, the true from the false, and the valid from the invalid.

(2) *The inevitability and inevitable experience of constraint (IIEC):* For empirical reasons, it has also been concluded that, even though IIE implies that all (normative, epistemological or empirical) standpoints are relative, they are not arbitrary. The situation both relativizes and constrains cognition and value judgements.

IIEC is sustained by scientific evidence but its fortune is also suggested by present and personal experience (i.e., on phenomenological grounds). This is, I can safely affirm, that at the present moment I am cognitively constrained. In regards to all (off-line and on-line) cognitive matters, my certainty is limited by my doubt, and the relationship between these two is not (directly) mediated by my own will. And so, when I compute 2 + 2, I can state without fear of dishonesty that I find myself constrained to provide 4 as an answer. In other words, I find 4 as a *necessary* answer.

If I claimed, however, that the characteristics of my experience of constraint are universal, in the sense that they apply to all, I would then fall into a phenomenological transcendentalism of sorts, and by doing so, would then find myself in a position at odds with IIE (which I find as *necessary* as 2 + 2 = 4).

It would then seem that the necessary outcome of this philosophical conundrum would be some sort of (normative) solipsism. This position, however, would also be at odds with IIEC. That is, even if I do not find the warrant to claim anything beyond my perspective at this present moment, the constraint I experience from my un-arbitrary perspective includes emotional and physiological necessities that I feel required to fulfil. As part of this required fulfilment I find myself in the need to negotiate with other aspects of my experience that seem to have the same characteristics as me (i.e., other people). For example, at this very present moment, one of the reasons I am writing this is the need to procure a safer financial future for my family and myself. I am, amongst other things, constrained to treat you, my reader, as precisely that: A reader. Finally, because I am constrained to IIEC, I am limited, cognitively exhausted, as regards to other ways in which the cognitive structure of those that I find similar to myself might work. This constraint, then, leads me far from solipsism, at least in practical or pragmatic terms.

The only outcome then that I am able (and therefore constrained and necessitated) to find for IIEC + IIE is the following: Because I cannot claim IIEC universally on account of IIE, and because I cannot imagine a situation different from IIEC in others like me: I am *constrained to hypothesize* that all others are also IIEC, but because of IIE *I must maintain myself open to the possibility of revision (scrutiny).*

The (situational) confirmation of my hypothesis can only be realized by my receiver upon communication. In this particular case, it can only be confirmed by you (my reader).

If this takes place, however, you will also find yourself incapable of finding a solution to the inscrutability of the impossibly observed, whilst at the same time you find yourself incapable of directly commanding the limits of your doubt and your certainty: *My account would then have categorical normative implications for both of us.* This is situational necessity.

The above example allows for an understanding of how situated necessity is able to provide validity of itself, without falling into any sort of transcendentalism. Situational necessity, however, is able to do this for any intellectual position. For example, I can safely state without fear of dishonesty that I am constrained to find the statement "All bachelors are unmarried" as necessarily being unproblematic. This does not imply that I find it "true" in a metaphysical sense (again, if so I would then do violence to IIE, which I also find as necessarily unproblematic). It rather implies that in this precise moment and from my perspective I have no option—I am cognitively exhausted—but to find it unproblematic in the same way that I find 4 to be the only result for the computation $2 + 2$. This implies that I am also constrained to hypothesize that you, my reader, are as constrained as I am to find it unproblematic. I am open, however, to the possibility that upon further knowledge—that is, upon future communication and experience—my limit of certainty on the matter may be altered. It may be the case that it is you who will change my limit of certainty and, therefore, you do not find it as unproblematic as I do. I am unable to know what sort of communication or experience could alter my limit of certainty in this regard. If I knew, it would mean that my limit is already altered. If, however, it were the case that upon communication with another agent I realize something (which I cannot foresee) that enables me to understand a problem in $2 + 2 = 4$, my cognitive capacity (we could call this "cognitive affordance") in this regard would be dramatically changed as a result.

Furthermore, because such a change takes off precisely from my previous position, it therefore contains it. In other words, because the previous state served as the ground for the change at hand, the resulting outcome must then be judged as superior or complementary from the perspective of the previous situation. The new, and necessarily previously unforeseen, possibility of action or description (affordance) is, from the perspective of the past cognitive situation, seen as better or complementary. *From the perspective of the affected agent, all changes to limits of certainty are cognitive enhancements. They are positive.*

In this regard, from the perspective of situated necessity, the primary standard of epistemological valuation is not the presently identified *form* which seems to constitute my cognitive exhaustion and thereby the means of the current cognitive constraint (i.e., my present reading of the identified rules of logic and/or fortunate induction) since, from the perspective of situated necessity, this identified *form*, its rules or characteristics, is always subject to change. Principles of deduction and induction can be understood as first person readings of phenomenological processes that lead to constraint. These, however, can only provide explanatory import. If there is any individual epistemic standard that can be obtained from the present account, it is *cognitive change itself*. A key epistemological principle that I am exhausted/constrained to identify (and exhausted/constrained to hypothesize you will recognize) is that, in epistemological terms, *the future is stronger than the present and the unknown stronger than the known*.

This does lead to a (situationally) necessary understanding of epistemic value judgement. Epistemic value is (involuntarily) awarded by an agent to that which changes his/her limits of certainty. From the perspective of a determined agent's cognitive situation, however, not all communication or experience will result in changes to his/her limits of certainty. For example, in an hypothetical exchange between agent (a) and agent (b), if (b)'s interpretation of (a)'s communication proposed the existence of the Loch Ness monster because of a picture (a) had seen in a sensationalist journal, it may be the case that (b)'s limits of certainty in regards to the fauna of Loch Ness will not be altered in any significant way. If (a), however, shows (b) the documentation of an HIV vaccine testing that she (a) co-ordinated in which said prophylactic proved 100 % effective in 1000 human trials, then the limits of doubt and certainty that (b) held in regards to the possibilities of AIDS prevention would be altered. That is, in relation to the *time/moment before the exchange* (TMBE), (b) will now experience either a higher degree of doubt or a higher degree of certainty regarding the matter at hand. From her post-TMBE cognitive perspective, (a)'s communication was laden with epistemic validity.

This epistemic standard, however, does not allow for an individualist epistemology. Amongst other things, the knowledge that results is one in which knowledge as a natural phenomenon cannot be reduced in individualist terms: Not only are all accounts of what exactly causes changes (i.e., rules of logic and induction) subject to change themselves, but also whilst situational necessity hypothesizes a natural understanding of situated epistemic validity (that which caused cognitive change), it does so in a way that makes it unable for this to be proven in any reliable way. This is because it is only the affected agent (b) who is aware of the situation and, furthermore, always able to choose not to disclose the validity that his own position is awarding to a specific position.

From this understanding of situated epistemic validity, however, a number of other premises necessarily obtain. These constitute a specific understanding of the relationship between social dynamics, cognitive hierarchy, and epistemic scrutiny. It is from these that hypothetical courses of action regarding the organization of knowledge production activities can be developed.

We have already explained (1) IIE and (2) IIEC, which as we have seen lead to:

(3) From the perspective of the affected agent (b), all cognitive change is positive. From the perspective of the agent, all cognitive change translates into cognitive enhancement.

This, in turn leads to:

(4) Because (2, IIEC) cognitive change is not subject to an agent's volition and because (3) (from the perspective of the affected agent) all cognitive change is positive, then if a certain agent (a) is able to change the limits of certainty of another agent (b) through communication, this would mean that agent's (b) cognitive situation awards (without the need of [b]'s volition) cognitive authority to (a)'s communication.

This last concept (4) can be fortunately called *natural cognitive authority*, namely,

the capacity of a determined agent to change—through communication—the cognitive situation (limit of certainty) of another.

The notion of natural authority corresponds to a social rendering of the already described notion of situated epistemic validity that results from situated necessity. Natural authority, in turn, provides the framework for a naturalist and horizontal rendering of normativity.

Unlike other attempts at normative accounts, the source of authority (in this case cognitive/epistemic) does not transcend situation. From the perspective of the offered theory, *within an exchange between two agents (a and b) cognitive (or epistemic) authority always stems from the cognitive situation of the receiver (b)*. One (a) has cognitive authority over another agent (b) (and therefore one's statements hold epistemic validity) but only if one (a) is capable of changing agent's (b) limit of certainty.

In these terms, the historical validity of science can then be explained, not in relation to the approximation of a transcendent standard (i.e., science as true or universally valid) but rather on its incontestable capacity to change the limits of certainty within a specific historical context and thus change the way a great variety of cognitive agents understand the world.

The next hypothesis that obtains from the premises of situational necessity could be called the hypothesis of *the primacy of difference*. That is, as implied above, in order for a determined agent (a) to have natural authority over another (b), these must perceive some sort of difference between their positions, either because (b) perceives (a)'s as *adding* to his position, or because they see their positions as *conflicting* or *incommensurate*. Communication between two agents whose limits of certainty are perceived as similar is "easier" but does not promise (for either of them) the level of cognitive alteration (and therefore enhancement) that the communication between two agents in extremely different cognitive situations may produce. Experimentation (scientific or otherwise) in these terms can then be understood as the pursuit of knowledge by means of locating oneself, through the attainment of experience, in a different cognitive situation to one's peers.

In order to simplify matters, we have so far explained the theory of natural (cognitive) authority and the hypothetical cognitive exchange between two different agents as if only one of the agents (a) possessed cognitive authority over the other (b). It is more fortunate, however, to consider that each of two different agents who meet wants to have cognitive authority over the other, even regarding a single matter. One (a) will, almost certainly have a greater capacity to change the limits of certainty than the other (b). But also almost certainly, the other (b) will have the capacity to change the limits of certainty of (a). Any exchange will cognitively alter and therefore enhance all the agents. If time is not an issue, then the greater the difference between them, the greater the possibilities of cognitive enhancement.

The primacy of difference then can be stated as follows:

(5) In order for a cognitive agent (a) to change or have the possibility to change another cognitive agent (b)'s limit of certainty, his/her (a) cognitive situation must differ from the other's (b). The greater the difference, the greater the possibility for change. And since (3) all change is positive from the perspective of the affected cognitive agent (b), the greater the difference, the greater the promise of cognitive enhancement.

This understanding of cognitive authority (and epistemic validity), however, also has implications for the relationship between (social or cognitive) hierarchy and knowledge. That is, (4) above implies that the notion of natural authority takes into account a form of (temporal and metaphysically horizontal) hierarchy: In a relationship of (cognitive) natural authority, he who knows, namely he (a) who has the capacity to change the limit of certainty of another (b) sits in a higher cognitive position in regards to that particular relationship.

This hierarchy, however, dissolves as a result of its practice. The capacity to change limits of certainty is practiced through communication. Successful communication will result in both change and approximation (elimination of difference) between the states of both the agents who are part of the exchange. As (b)'s limits of certainty are changed and his cognitive situation is enhanced by communication, (a) begins to lose the capacity to change them. Hypothetically, if their communication is successful (this may take time or even training), (b) will have the same capacity as (a). As a result, (a) will lose all natural authority she has in relation to (b).

This leads to the fifth necessary hypothesis that obtains from IIE + IIEC:

(6) Cognitive authority (4), namely the capacity to change the limits of certainty of another agent and the hierarchical relationships that it implies are levelled/stabilized as a result of its practice (communication). In other words, the communication of knowledge and the resulting cognitive enhancement eliminate both cognitive difference and cognitive hierarchy.

Without any situation-independent (transcendent) standard that may provide capacity to determine or judge epistemic value, from the above four hypotheses—which (necessarily) obtain (in regards to my present situation and hypothetically in yours) from IIE + IIEC—one is constrained to predict the following:

All things being equal, which in this case means assume specifically:

X. that there are no major temporal constraints or boundaries for communication,
Y. that there are no major spatial or material constraints for communication, and
Z. that all participating agents are willing and incentivized to engage in communication until stabilization, and to acknowledge all changes that occur to their limits of certainty; then:

It can be predicted that unbounded communication will lead those with less capacity to change the limits of certainty towards cognitive enhancement, and that all positions (or aspects of positions) that survive change correspond to the highest level of cognitive enhancement for both these individuals and for the community. In so many words, according to the above necessary hypotheses, it can be predicted that, *ceteris paribus*, unbounded communication will act as a naturalist, specifically socio-cognitive, exercise of falsification or refutation, whilst at the same time also acting as a distributor of knowledge, and a levelling of cognitive hierarchy. Without any transcendent or objective standard it is only the difference, combined with constraint, that provides grounds for epistemic scrutiny.

Once all differences have been levelled, or levelled as much as the circumstances of its members may allow, the only way to acquire further collective enhancement is by finding a new difference.

The above situation serves as an ideal for the most effective circumstance for the acquisition of individual and collective cognitive enhancement. It is clear to see, however, what it is that gets in the way of this "open society", namely (X) temporal, (Y) spatial and material constraints, and—under the assumption that ranking members within a hierarchy has the incentive to maintain this position— then also (Z) the social boundaries for communication that a cognitive hierarchy is afforded to maintain in lieu of (X) and (Y). Or in other words, (Z) any social structure that might actively repress or set boundaries for communication that has the capacity to change the limits of certainty, or that might prescribe the individual psychological repression of these.

In lieu of the above, epistemology and/or philosophy of science would cease to be endeavors aimed at the prescription of fortunate means of knowledge production, or to be attempts to establish firm standpoints for epistemic judgement or demarcation. These philosophical efforts would now relocate themselves within the social and cognitive sciences in order to attempt to understand what form of social organization would provide the most un-intermediated, unbounded and efficient form of communication amongst the largest quantity of cognitive agents taking into consideration that which is afforded by the interplay between technology, spatial, material and temporal constraints, and the political incentives that may be in the way; an enterprise whose overall prescriptive spirit is not dissimilar to that of the NSE.

And so whilst strict naturalism may not be able to provide an epistemology, it is able to provide the means of a normative social theory of epistemological import, or in other words, it is able to provide a social epistemology.

References

Alston WP (1988) The deontological conception of epistemic justification. In: Tomberlin JE (ed) Philosophical perspectives, vol. 2: Epistemology. Ridgeview, Atascadero, CA, pp 257–299
Alston WP (1996) Belief acceptance and religious faith. In: Jordan J, Howard-Snyder D (eds) Faith, freedom, and rationality: philosophy of religion today. Rowman and Littlefield, New York, pp 3–27

Armstrong DM (2010) Introduction. In: Armstrong DM (ed) Sketch for a systematic metaphysics. Oxford University Press, Oxford, pp 1–5

Barnes B, Bloor D (1982) Relativism, rationalism, and the sociology of knowledge. In: Hollis M, Lukes S (eds) Rationality and relativism. Blackwell, Oxford, pp 21–47

Barnes B, Bloor D, Henry J (1996) Scientific knowledge: a sociological analysis. Athlone, London

Barsalou LW (1999) Perceptual symbol systems. Behav Brain Sci 22:577–600

Barsalou LW (2008) Grounded cognition. Annu Rev Psychol 59:617–645

Bernstein D, Jackson JD (2004) The Daubert trilogy in the states. Jurimetrics 44:1–17

Bowmaker JK (1980) Colour vision in birds and the role of oil droplets. Trends Neurosci 3(8):196–199

Brown PK, Wald G (1964) Visual pigments in single rods and cones of the human retina. Science 144:45–52

Camic C, Gross N (2001) The new sociology of ideas. In: Blau JR (ed) Blackwell companion to sociology. Blackwell, Malden, pp 236–249

Chemero A (2009) Radical embodied cognitive science. MIT Press, Cambridge, MA

De Caro M, Macarthur D (2004) Naturalism in question. Harvard University Press, Cambridge

De Caro M, Macarthur D (2010) Naturalism and normativity. Columbia University Press, New York

Dennett DC (1978) Toward a cognitive theory of consciousness. In: Dennett DC (ed) Brainstorms: philosophical essays on mind and psychology. MIT Press, Cambridge, MA, pp 149–173

Fleck L (1979 [1935]) Genesis and development of a scientific fact. University of Chicago Press, Chicago

Fodor J (1985) Fodor's guide to mental representations: the intelligent auntie's vadecum. Mind 94:76–100

Fodor J (1987) Psychosemantics: the problem of meaning in the philosophy of mind. MIT Press, Cambridge, MA

Fuller S (1997) Science. Open University Press, Buckingham

Fuller S (2000) The governance of science: ideology and the future of the open society. Open University Press, Buckingham

Fuller S (2002) Social epistemology, 2nd edn. Indiana University Press, Indianapolis

Fuller S (2007a) New frontiers in science and technology studies. Polity Press, Cambridge

Fuller S (2007b) The knowledge book. Acumen Publishing, Stocksfield

Fuller S (2009) The sociology of intellectual life. Sage, London

Fuller S, Collier JH (2003) Philosophy, rhetoric, and the end of knowledge: a new beginning for science and technology studies, 2nd edn. Lawrence Erlbaum Associates, Publishers, London

Gallaguer S (2005) How the body shapes the mind. Oxford University Press, Oxford

Gallaguer S, Zahavi D (2008) The phenomenological mind: an introduction to philosophy of mind and cognitive science. Routledge, New York

Gibbs R Jr (2003) Embodied experience and linguistic meaning. Brain Lang 84:1–15

Goldman A (1993) Philosophical applications of cognitive science. Westview Press, New York

Goldsmith TH (1990) Optimization, constraint, and history in the evolution of eyes. Q Rev Biol 65:281–322

Hofstadter D (1995) Fluid concepts and creative analogies: computer models of the fundamental mechanisms of thought. Basic Books, New York

Hume D (1888 [1740]) A treatise of human nature (edited by Selby-Bigge LA). Clarendon Press, Oxford

Hurvich LM, Jameson D (1957) An opponent theory of colour vision. Psychol Rev 64: 384–404

Hylton P (2007) Quine. Routledge, New York

Kiefer M, Barsalou L (2013) Grounding the human conceptual system in perception, action, and internal states. In: Prinz W, Beisert M, Arvid H (eds) Action science: foundations of an emerging discipline. MIT Press, Cambridge, MA, pp 381–407

Kiefer M, Sim EJ, Hermberger B, Grothe J, Hoenig K (2008) The sound of concepts: four markers for a link between auditory and conceptual brain systems. J Neurosci 28: 12224–12230

Kitcher P (2001) Science, truth, and democracy. Oxford University Press, Oxford

Knorr-Cetina K (1999) Epistemic cultures: how the sciences make knowledge. Harvard University Press, Cambridge, MA

Kornblith H (1994) Naturalism: both metaphysical and epistemological. Midwest Stud Philos 19(1):39–52

Kuhn T (1970) The structure of scientific revolutions. University of Chicago Press, Chicago

Lakoff G, Johnson M (1999) Philosophy in the flesh: the embodied mind and its challenge to western thought. Basic Books, New York

Lakoff G, Nuñez RE (2000) Where mathematics comes from: how the embodied mind brings mathematics into being. Basic Books, New York

Latour B (1988) The pasteurization of France. Harvard University Press, Cambridge, MA

Latour B, Woolgar S (1986) Laboratory life: the construction of scientific facts, 2nd edn. Princeton University Press, Princeton

Lochhead J (1995) Hearing new music: pedagogy from a phenomenological perspective. Philos Music Educ Rev 3(1):34–42

Longino HH (2002) The fate of knowledge. Princeton University Press, Princeton

Mannheim K (1929) Ideology and utopia: an introduction to the sociology of knowledge (trans: Wirth L, Shils E). Harcourt, Brace and Company, New York

Matthen M (2005) Seeing, doing, and knowing. Clarendon Press, Oxford

Melnyk A (2003) A physicalist manifesto. Cambridge University Press, Cambridge, MA

Merton RK (1970) The sociology of science. University of Chicago Press, Chicago

Nagel T (1986) The view from nowhere. Oxford University Press, Oxford

Parsons T, Platt G (1973) The American university. Harvard University Press, Cambridge, MA

Pulvermüller F (2005) Brain mechanisms linking language and action. Nat Rev Neurosci 11:351–360

Putnam H (1982) Why reason can't be naturalized. Synthese 52:3–23

Putnam H (2010) Science and philosophy. In: De Caro M, Macarthur D (2010) Naturalism and normativity. Columbia University Press, New York, pp 89–99

Pylyshyn Z (1984) Computation and cognition: toward a foundation of cognitive science. MIT Press, Cambridge, MA

Quine WVO (1951) Two dogmas of empiricism. Philos Rev 60:20–43

Quine WVO (1960) Word and object. MIT Press, Cambridge, MA

Quine WVO (1969a) Ontological relativity. In: Quine WVO (ed) Ontological relativity and other essays. Columbia University Press, New York, pp 27–68

Quine WVO (1969b) Epistemology naturalized. In: Quine WVO (ed) Ontological relativity and other essays. Columbia University Press, New York, pp 69–90

Quine WVO (1975) On empirically equivalent systems of the world. Erkenntis 9:313–328

Quine WVO (1981) Theories and things. Harvard University Press, Cambridge, MA

Ritchie J (2008) Understanding naturalism. Acumen, Stocksfield

Rouse J (1987) Knowledge and power: towards a political philosophy of science. Cornell University Press, Cornell

Rudder Baker L (2013) Naturalism and the first person perspective. Oxford University Press, Oxford

Sellars W (1963) Philosophy and the scientific image of man. Sci Percept Reality 2:35–78

Skarda CA, Freeman WJ (1987) How brains make chaos in order to make sense of the world. Behav Brain Sci 10:161–195

Spinoza B (1996 [1677]) Ethics (edited and translated by E. Curley). Penguin Books, London

Thelen E, Smith L (1994) A dynamical systems approach to the development of cognition and action. MIT Press, Cambridge, MA

Thompson E (1995) Colour vision: a study in cognitive science and the philosophy of perception. Routledge, London

Thompson E (2007) Mind in life: biology. Phenomenology and the sciences of the mind. The Belknap Press of Harvard University Press, Cambridge, MA

Thompson E, Varela F (2001) Radical embodiment: neural dynamics and consciousness. Trends Cogn Sci 5:418–425

Van Gelder T (1995) What might cognition be, if not computation? The Journal of Philosophy 92(7):345–381

Varela FJ, Thompson E, Rosch E (1991) The embodied mind: cognitive science and human experience. MIT Press, Cambridge, MA

von Helmholtz H (1970 [1866]) Physiological optics—the sensations of vision (translated). In: MacAdam DL (ed) Sources of color science. MIT Press Cambridge, MA., pp 84–100

Wessel D (2006) An enactive approach to computer music performance. In: Orlarey Y (ed) Le feedback dans la création musicale. Lyon, France, pp 93–98

Young T (1802) The bakerian lecture: on the theory of light and colours. Philos Trans R Soc Lond 92:12–48

Prolegomena to a Genealogy of the Transgressive Mindset

Steve Fuller

Abstract

This paper is divided into two parts, one on the theological and the other on the philosophical roots of the transgressive mindset, which when seen against the backdrop of Western intellectual history is ultimately about establishing a continuity of being between the human and the divine. Along the way, various intermediate positions are discussed, including Prometheus, Faust, Superman, and Plato's philosopher-king—with the image of Jesus figuring in the background as an unstable human-divine hybrid. The linchpin philosopher for consideration of the transgressive mindset turns out to be the medieval scholastic John Duns Scotus, whose theory of "univocal predication" underwrites modern notions of literalness, which opens up the prospect of converting possibility into actuality through an act of will, as exemplified by the ontological argument for the existence of God, in which the deity self-realizes by definition.

Keywords

Prometheus · Faust · Superman · Philosopher-king · Cyborg · John Duns Scotus

This article is dedicated to my student Morteza Hashemi Madani, whose Ph.D. on the religious roots of contemporary atheism I supervised while thinking deeply about this topic.

S. Fuller (✉)
Coventry, UK
e-mail: s.w.fuller@warwick.ac.uk

© Springer Fachmedien Wiesbaden GmbH 2016
B.-J. Krings et al. (eds.), *Scientific Knowledge and the Transgression of Boundaries,* Technikzukünfte, Wissenschaft und Gesellschaft / Futures of Technology, Science and Society, DOI 10.1007/978-3-658-14449-4_11

In what follows, I consider successively the theological and the philosophical roots of the transgressive mindset, which in the Western tradition has been about establishing a continuity of being between the human and the divine. Along the way, I examine various intermediate positions, including Prometheus, Faust, Superman, and Plato's philosopher-king—with the image of Jesus figuring in the background as an unstable human-divine hybrid. The medieval scholastic John Duns Scotus plays a pivotal role in my account, mainly for his theory of "univocal predication", which underwrites modern notions of literalness. Specifically, Duns Scotus opens up the prospect of converting possibility into actuality through an act of will, as exemplified by the ontological argument for the existence of God, in which the deity self-realizes by definition.

1 Back from Prometheus to Faust and Simon Magus: The Theological Roots of Transgression

Blumenberg (1985) may well be correct to assert that in today's world the myth of Prometheus serves as the template for humanity's ambitions to transcend its intermediate status between animal and deity. Certainly the myth is a transgressive one. Prometheus, an angel-like figure, steals fire from the Greek gods and gives it to humanity, for which he himself is then punished and humanity is consigned to an uncertain fate. In Christian theology, the closest analogue is Satan, a somewhat shadowy figure in the Bible who is given clear form in the early modern era via John Milton's *Paradise Lost*. But this shift in mythic context also suggests that the more appropriate model for the transgressive being is Adam, the first human, who is tempted by Satan's emissary to eat of the forbidden fruit, which hangs from the tree of knowledge of good and evil (Genesis 2: 16–17). In that case, *Faust*—not Prometheus—is the better starting point for our discussion. Faust is invariably portrayed as a human—not an angel—who in some sense oversteps his divine entitlement, typically in ways that implicate the interference of his animal nature (i.e., a woman is involved). The main post-Adamic Biblical precedent for Faust is Simon Magus, a very interesting figure who appears in the Acts of the Apostles in the New Testament—more about whom below. Indeed, Prometheus only starts to compete seriously with Faust as an icon of humanity's transgressive tendencies with the English translation of the ancient Greek classics in the early nineteenth century, an important beneficiary of which was Mary Shelley, who subtitled her 1818 novel on Frankenstein "The Modern Prometheus" and began with a quote from *Paradise Lost*.

But before turning to the image of Faust as transgressor, it is worth observing that Prometheus has at least two rivals for the Greek mythic source of transgressive humanity: Tantalus and Sisyphus. Tantalus was a son of Zeus by a human mother who doubted the omniscience of the gods and so put it to the test by seeing whether they would recognize Tantalus' own son served up as the main course of a meal. Zeus saw through the ruse and condemned Tantalus to stand in a pool of water in Hades surrounded by sources of nourishment that receded whenever he approached them. The punishment may be understood as apt for someone who underestimates the distance between the gods and their creations. In contrast, Sisyphus, the founding king of Corinth, enjoyed disobeying and deceiving the gods as he extended his rule. For his insolence, he was sentenced to push a boulder up a hill that only rolled back down whenever he approached the top. This punishment is meant to signify the ultimate vanity of power pursued for its own sake. It merely consumes the energies of those in its pursuit, as Sisyphus forever fails to see that all his efforts constitute a "sunk cost" that will never be redeemed. Taken together, these two myths suggest a picture of humans failing to realize that there is more to divinity than simply the indefinite extension of human capacities: It also involves orienting those capacities to the right ends. It is here that Prometheus poses a special challenge because his tale is less about doubting the power of the gods per se than doubting that such power is exclusively theirs.

Prometheus' original crime lay in tricking Zeus into accepting a superficially pleasing sacrifice, in response to which Zeus withdrew fire from humans, which then Prometheus promptly stole back for them. He was punished for this second malfeasance by his liver being eaten by an eagle, only for the organ to be regenerated and eaten once again. It is the sort of punishment that only a god could suffer in perpetuity—never to be relieved in outright death—since only gods have the power of self-regeneration. (If transhumanists get their way, and we have indefinite lifespans, then this might come to replace capital punishment.) Transferred to a Christian context, the Prometheus narrative suggests that God deals with humans on a discretionary basis, at first enabling them to share divine properties but then withdrawing that capacity in light of some transgression, say, as described in Adam's Fall. In that case, the presence of a mediator who might circumvent divine will is cast as the ultimate evil who threatens to undo the specific order imposed by God. Thus, Milton's Satan is a being whose power grows with the onset of "disorder", once divinely circumscribed capacities in the human are allowed to run amok, as in the case of an insatiable curiosity or an unquenchable thirst for power.

In contemporary transhumanist terms, Prometheus and Satan are purveyors of an extreme version of "ableism", the ideology of the indefinite expansion of particular human capacities, even if that entails exploding the integrity of the person, understood as a bundle of mutually limiting capacities that are jointly focused on realizing some reasonable life project (Wolbring 2008). It is perhaps easiest to envisage this prospect in materialist terms, namely, as "living within one's means", e.g., that the frontal lobe of the cerebral cortex cannot expand indefinitely without making the human body structurally unstable—hence the physical awkwardness associated with Frankenstein and associated cyborg creations in science fiction. Against this backdrop, God is presumed to have designed the body in just the right proportions to enable it to function as it should.

At this point, I want to leave the discussion of Prometheus/Satan because, in the end, both narratives portray humans as pawns in an eternal chess match between Good (aka Zeus/God) and Evil (aka Prometheus/Satan). To be sure, this image is familiar from the cosmologies of Zoroastrianism. and Manichaeism, the latter often treated as a Christian heresy. However, the Faust tale more directly implicates humans in their own predicament, as they actively seek to appropriate what they believe is rightly theirs. In this context, there is little for Evil to do other than nudge humans to do something that they are already predisposed to do. Satan's emissary, Mephistopheles, presents the matter as a bargain, in which Faust is forced to admit that there is a cost for transgressing the divine settlement with Adam (i.e., his descendants are fallen but they are promised redemption, in a manner to be determined by God). Strictly speaking, the cost to Faust involves the acceptance of risk, since he cannot foresee all the consequences of his pact with the devil, as opposed to God's own unilaterally imposed securitized arrangements for human salvation.

The modern image of Faust underwent a subtle metamorphosis from the dramatic portrayal offered by Christopher Marlowe in the early seventeenth century to that by Johann Wolfgang von Goethe two centuries later. Whereas Marlowe's Dr. Faustus meets a grisly end after refusing to repent his ways, Goethe's Faust receives divine salvation despite his erring ways. The *Ur*-text of the tale is arguably Book V of Augustine's *Confessions*, where someone named "Faustus" is presented as incredibly learned yet ignorant of his own limitations. For Augustine, Faust's arrogance epitomized our inherited fallen state, even as we have been created "in the image and likeness of God" (*in imago dei*), a Genesis formulation to which Augustine himself had drawn significant hermeneutical attention (van der Laan 2007, p. 7). From Augustine's time to the start of the Protestant Reformation, this turn of mind had been associated mainly with the Pelagian heresy, which invested in humans the capacity to redeem themselves from Original Sin without any divine mediation, simply by virtue of possessing a will to do so, albeit the very same will that first led Adam down the path of sin (Passmore 1970, Chap. 5). However, alongside—and of equal significance by the time of the Scientific Revolution—was the Arian heresy, which questioned the uniqueness of Jesus as a human invested with divine powers of redemption (Fuller and Lipinska 2014, Chap. 2).

The difference is worth highlighting in terms of the transgression that might be attributed to a Faust-like figure: The Pelagian supposes that humans can find their own means to achieve divine ends, while the Arian supposes that humans "always already" possess divine capacities which may have yet to be discovered. In terms of contemporary philosophy of science, one might associate the Pelagian with a "constructivist" and the Arian with a "realist" attitude towards human nature—or, more precisely, humanity's divine nature. In that case, Marlowe's Dr. Faustus is a failed Pelagian, and Goethe's Faust a failed Arian. The significance of "erring" in the two cases is different. In Marlowe's case, it reflects a bloody minded sense of one's own correctness, whereas in Goethe's case it reflects a sense of infinite striving to achieve an ideal state: The one is condemned, the other rewarded (van der Laan 2007, Chap. 10). The relevance of this difference to the Scientific Revolution is that taboo "magical" practices like alchemy come to be seen in a different light, independently of any change in their actual efficacy, again a shift from a constructivist to a realist interpretation. Thus, instead of imagining alchemists as constructing fraudulent versions of effects that only God could produce, they come to be seen as trying—often unsuccessfully—to imitate divine agency in their own practice. A frequently cited Renaissance precedent for Faust, the sixteenth century philosopher-physician Paracelsus, regarded the experimental arts as micro-expressions of divine creativity and contributed to what is nowadays regarded as the "hermetic" roots of the Scientific Revolution (Yates 1964, Chap. 8).

To be sure, from today's standpoint, the Renaissance turn to realism looks like a salutary shift towards a modern understanding of practitioners of the magical arts. What previously might have appeared to be acts of deception or self-deception now looked like sincere inquiry. However, at the time the shift made the magicians seem still more dangerous to clerical authorities, eventuating in the executions of, say, Giordano Bruno and Michael Servetus. A focal point for this anxiety was the Socinian heresy, which made its way in the sixteenth century from Italy to Jagiellonian Poland, the most liberal Christian regime of the time (e.g., as home to Copernicus), where the heretics were known as the "Polish Brethren". Their most notable member was one Fausto Sozzini. Perhaps the most distinctive feature of their legacy was its ambiguous take on Arianism, which shaped the Enlightenment's theological tendencies to Deism and Unitarianism, especially through John Locke's ideas of a "reasonable Christianity" (Wallace 1984). Socinians clearly denied the *uniqueness* of Jesus' divinity. But was this because everyone is born divine or no one is? In the former case, our species task is to unleash divine powers that we each possess individually, which is a counsel of liberty; in the latter, our species task is to achieve the divinity we lack as individuals by means of some form of collective self-organization, a counsel of solidarity. The full range of modern political ideologies from libertarianism to communism are licensed under this rubric, though the version that has proved most durable is the balanced perspective epitomized in the US Constitution, a document notable in its day for making human self-realization—not simply domestication—a central task of government, one which in the past might have been left to churches (Hillar 2009).

The person who brought all of the above strands of thought into focus for the image of Faust that appears in Goethe was Germany's answer to Voltaire, Gotthold Lessing. Lessing was widely known to have written, but never published, the first notable version of the Faust tale which ends with protagonist being saved, not damned (van der Laan 2010). Lessing's Faust is an internally divided yet ultimately sympathetic character. On the one hand, he is sufficiently arrogant to interpret his having been created "in the image and likeness of God" quite literally to mean that he need not adopt the supplicant's position in prayer, the point of which would be to have the deity look kindly on a request without questioning the divine *modus operandi*. Rather, the Faustian communicates with God on the presumption of equality by means of his own acts of technical creation and interrogating the divine *modus operandi* via testable hypotheses. Yet, on the other hand, Faust is, in the end, a mere human, who is constitutionally prone to error, including in terms of his capacity to learn properly from error—which is tantamount to learning from experience, since every experience is potentially an error if it is not responded to properly. This meta-level sense of fallibility (including a reference to Goethe) can be found in Norbert Wiener's concerns about the perils of "positive feedback" in cybernetic systems (Wiener 1961, p. 176). I shall return to Wiener below, but it is worth mentioning that the prospect of harnessing the redemptive power of error informs Karl Popper's Goethe-inspired observation that humans began to make rapid intellectual progress once we learned to detach ourselves from the divine *modus operandi* (Popper 1972).

Integral to the Faustian turn is the violation of a specific taboo whose observance enabled a world order as comprehensive and stable as that of Roman Catholic Christendom to survive as long as it has. Put in today's terms: *Qualitative differences cannot be converted to quantitative ones simply by an act of human will.* (In the next section, I explore this point more deeply under the rubric of "taboo cognition".) In whatever sense humans may have been created *in imago dei*, they cannot measure the immeasurable or compare the incomparable, which includes *inter alia* creating life from non-life, not least the "homunculus", something which virtually all versions of Faust, both in fact and fiction, are accused of at some point (Ball 2011). This doctrine, familiar in pagan culture from Aristotle, is nowadays discussed in terms of "irreducible incommensurability" or, more simply, a "patchwork reality" (e.g., Galison and Stump 1996). The underlying assumption is that if there is an ultimate deity, the logic of its creativity is impenetrable to humans and hence we are forced to experience reality's intelligent design in an unresolved fashion. Thus, any claims to having fathomed a common set of a few underlying principles for generating the full diversity of observable phenomena—of the sort that physicists continue to search for under the rubric of "Grand Unified Theories of Everything"—was seen as *prima facie* heretical. This helps to explain Isaac Newton's own caginess with regard to whether his "laws of nature" represented mere empirical regularities or God's original recipe for Creation. To be sure, Newton's own reticence did not stop John Maynard Keynes from dubbing him "The Last Magician", admittedly from the safe secular distance provided by the 300th anniversary of Newton's birth.

This is an apt moment to consider the figure of Simon Magus, a Biblical source for Faust and arguably the patron saint of all heretics. The crucial Biblical episode (Acts 8: 9–24) was instrumental in the coinage of "simony", the buying and selling of church services, a symbol of corruption that was frequently cited by the early Protestant Reformers as grounds for splitting with the Church of Rome. In the original story, which takes place after the death of Jesus, Simon is portrayed as a Samaritan divine and recent Christian convert who is very impressed by Peter's powers to conjure up the Holy Spirit. Thus, Simon asks him at what price those powers might be bought. Peter disdainfully responds that such powers are a gift and not for sale, and the very fact that Simon thinks otherwise demonstrates his unworthiness of the gift. The Biblical author—conventionally thought to be Luke of the Gospels—is probably reminding his readers of Adam's original attempt to arrogate for humanity a power that can only be provided by the Grace of God. In any case, clearly Simon had transgressed a sacred ontological boundary by attempting to render a difference in kind into one of degree—in this case, divine powers understood as a highly potent skill-set that nevertheless might be obtained via human currency.

Two aspects of this story are relevant to the development of the Faustian imagination. First is that divine powers might be copied, simulated or otherwise reproduced by human means. Second is that this capacity might be acquired by a transaction involving humans renegotiating their relationship with God. In other words, *contra* Prometheus, the Faustian does not attempt to steal divine powers or operate behind God's back. Rather, he sees the securing of divine powers as a reasonable request for someone who has already made the basic faith commitment to the Christian deity, and hence recognizes his fallen state. Interestingly, even in the authorized Acts version, Simon has the final word, hoping that God will not abide the curse that Peter had effectively cast on his soul. In the past century, perhaps the most interesting secular update of this sensibility was Thomas Mann's 1947 novel, *Doctor Faustus*, in which the main character, the composer Adrian Leverkühn, deliberately contracts syphilis to induce a madness that leads him (via an hallucinated exchange with Mephistopheles) to unprecedented levels of creativity, each exacting a greater toll on his body and soul, as if the entropy principle were in operation.

Peter, the Apostle whom Jesus anointed as his immediate successor, was not the only early church leader disturbed by Simon Magus. Several other stories involving Simon were edited out of the Bible in writings declared "apocryphal" by various ecclesiastical councils in Christianity's early history. Georg Wilhelm Friedrich Hegel's great theological follower, Ferdinand Christian Baur—the founder of the Tübingen critical-historical school—made the problem of Simon Magus explicit. In terms popularized by the institutional economist Williamson (1975), Simon was a man of the "market", whereas his nemesis Peter was a man of the "hierarchy"—at least when it came to the organizational logic of Christianity. For this reason, Baur argued that "Simon Magus" did not exist as such but was a euphemism for what Peter's followers despised about St Paul's missionary work, as exemplified in his Epistles, which basically involved selling Christianity to a variety of Mediterranean peoples in terms that enabled them to see the relevance of Christ's message to their lives, as Paul himself had, without having directly encountered the human Jesus or someone whose own authority descended from contact with Jesus. In other words, Paul proposed that it was possible to be a Christian simply by allowing the Word of God to circulate in one's own mind, a simple act of persuasion, the result of which would be a voluntary change of life that one would regard as an improvement.

But this meant that Paul presented the message of Jesus in different terms to different audiences, highlighting spiritual and practical conflicts to which each target audience would be already attuned. Such a strategy suggested that Paul thought of Christian conversion on the model of merchandising, in which the exact price is determined in terms of the target audience's need and budget. Thus, some audiences—e.g., Stoics—may find it easier to convert than others because they are already doing and thinking in ways that would make it easy for them to accommodate Christ's message. Indeed, this is how historical specialists on the spread of Christianity normally see the matter.

The problem arises when this retrospective explanation is seen as having been part of Paul's original *modus operandi*, which could be seen as undermining the qualitative change of being implied by the idea of "conversion", if not an outright expression of cynicism on Paul's part. After all, Paul could be seen as someone simply interested in increasing the number of Christian believers without much concern for the interpersonal consistency of their beliefs, very much like an advertiser or campaigner who does not really care how one comes to purchase a product or vote for a policy, just as long as they manage to "close the deal". In that case, it would be left to "Divine Grace", perhaps ironically understood, to determine whether the converted Christian received "value for money" at the Final Judgement. Arguably this is what Calvinism is all about.

However, this then raises a politically delicate historical paradox, which Max Weber and Werner Sombart debated at the dawn of the twentieth century vis-à-vis the responsibility of Protestants versus Jews for the rise of capitalism. The main practices that Christians would have had to model their individual negotiations of the faith in the name of a universal Christianity à la St Paul would have been the money-based exchange relations that were relegated to the Jews until the modern period. Indeed, the wealth that the Jews accumulated by brokering successful exchanges may have been the model for the secular success of a Christendom, understood as a Christian empire, closed under the common spiritual currency underwritten by the Church. However, the Jewish brokers themselves would be deemed sacrilegious because they assumed (without authorization) the divine position of personally determining the rate of exchange. This is the background against which it made sense to portray Satan's agent in the Faust legend, Mephistopheles, as Judaic in aspect.

2 The Indefinite Pursuit of Reason for Its Own Sake: The Philosophical Roots of Transgression

Thoughts about "the limits of rationality" normally spark images of things that get in the way of reason: Emotions, bias, lack of evidence, etc. We rarely think about cases in which reason falls short of its own potential, but that certainly happens. In this respect, reason becomes its own worst enemy by not trusting its own processes. Here I mean situations, always quite striking in retrospect (i.e., once we "return to reason"), when something that we had believed on independent, usually metaphysical, grounds prevented us from applying a line of argument to its logical conclusion.

Such arguments typically have a mathematical character. Specifically, they attempt to convert a difference in *kind* to one of *degree*. For example, you start by thinking that X is good but Y is bad, but then your friend observes that Y is really not so different from X, and so if you accept X you should also accept Y. In order to appear to future observers as having been on the right side of history, you should concede your friend's point. Put as a rule of thumb: In a dispute between someone who claims that A and B are radically different from each other and someone else who claims that A and B are versions of each other, the second person has history on his side and will be eventually regarded as the more progressive thinker.

Consider the following line of thought: If we can measure moving bodies in the heavens and on Earth, why can't we also measure the motions of people on Earth or, for that matter, the motions of whatever "spirit" lurks in their bodies? Until the late nineteenth century, the overriding objection to such projects—be it called "political economy" or "experimental psychology"—was that people are somehow qualitatively different from ordinary physical things. That metaphysical belief casts a pallor of moral corruption on anyone who would dare derive useful information from measuring or calculating the activity of human beings. It has proved to be the main retardant in the development of the social sciences. The attitude lingers today, say, in the instinctive antipathy displayed by even quite learned people to economic appraisals of value or behavioral indicators of thought. The US political psychologist Tetlock (2003) has coined the phrase *taboo cognition* for this self-limitation of reason.

The ancient Greek logicians may have made the original discovery of taboo cognition when they pondered the paradox surrounding the style of argumentation that we now call "slippery slope". A typical argument of this kind goes as follows: A full head of hair clearly makes one "hairy". But suppose just a single hair is removed: Is not that person still "hairy"? Of course. Then you remove another, and another, and so on. Clearly, at some point, the person is no longer "hairy" but "bald"—but when exactly? Over my career, I have had occasion to consider the various mechanisms deployed to transgress taboo cognition, from the Greeks to Tetlock himself (Fuller 1988, 2005, 2010b). The general lesson to be learned is that once an absolute distinction such as "hairy" versus "bald" is turned into a continuum admitting of degrees of "hairiness" or "baldness", then intuitions relating to the values carried by the distinction—such as the difference between age and youth—start to be lost. After all, a value can carry no normative force (i.e., be officially enforced) in a society unless there are reliable empirical indicators of the value's presence or absence. In that case, once one realizes that some relatively bald people are quite young and some relatively hairy people are quite old, the taboo distinguishing old and young has been breached, if not outright broken. In that case, those who still wish to mark a socially recognized distinction between "youth" and "age" need to re-draw the boundary on some other empirical basis.

If the values at stake in being hairy or bald seem to be of purely cosmetic interest, then consider more morally freighted taboo topics such as the difference between "living" and "non-living" or, for that matter, "human" and "non-human". Once we admit the existence of intermediate cases, the significance of these distinctions is similarly called into question. This helps to explain the heated controversies generated by, on the one hand, the prospect of what used to be imagined as "creating life in a test tube" but now goes by the no-nonsense name of "synthetic biology", and, on the other, the very idea that only a few genes separate humans from the other primates. Taken to the extreme, in full taboo violation, why not say that a rock is, in some sense, a "minimum" form of life or a human is, in some sense, a "maximum" form of matter? Here the word "complexity" is often invoked to capture the continuum along which the rock and the human lie at opposite ends, though in this context "complexity" does more metaphysical work than can be reasonably expected of it. Two self-styled "monistic" philosophies popular among scientists in the late nineteenth and early twentieth centuries—*panpsychism* (i.e., matter as unrealized mind) and *hylozoism* (i.e., mind as emergent from matter)— captured this spectrum of thought, which has persisted longest in the biological sciences (Weir 2012).

The more that science has accepted the atomic world view, which treats the qualitative differences between things as literally a surface appearance of an underlying combinatorial reality, the easier it has been to confront taboo cognitions. The overall effect of this intensive application of science has been to weaken the taboos, allowing reason to flow more freely. But it would be a mistake to think that science alone is capable of removing the barriers to our self-imposed restrictions on reason. In fact, *poetry* may be seen as anticipating science in its taboo-busting function.

The Greeks originally understood poetic activity as the production of worlds with words, with all the potential for fraud suggested by such a definition, as people tend to be moved by the striking images they hear or see, regardless of whether they correspond to anything in empirical reality. Thus, the Greeks seem to have had the concept of a hologram *avant la lettre*. At least Plato thought so, which led him to ban poets lacking official authorization from his ideal republic. Against this philosophical liability, Jesus Christ's youngest publicist, John the Evangelist, deserves credit for re-instating "creation by the word" (*logos*) to unequivocally divine status, which perhaps reached its fullest expression in the modern era through the Romantic movement (Bloom 1973). In any case, poetry's world production requires a rather disciplined use of the imagination in which *metaphor* plays a central role.

The cognitively striking feature of metaphor is its rendering of two seemingly quite different things very similar, if not two instances of the same underlying reality. When a metaphor works, it opens up a new way of seeing the world. And once a metaphor has been sufficiently elaborated, it turns into an analogy, on the basis of which models may be constructed for scientifically comprehending and sometimes even controlling the world. The Scientific Revolution of the seventeenth century marked a watershed in this intellectual trajectory, which amounts to taking figures of speech with an unprecedented degree of literalness, typically with the aid of mathematical reasoning (Hesse 1963). Of course, scientific models harbour their own mental restrictions. So for the past two centuries or more, certainly since Immanuel Kant's *Critique of Judgement*, there has been much fussing about the boundary between a *machine* and an *organism*, especially when the identity of the "human" has been involved. The elusive field of "biophysics", counterpoised to both artificial intelligence and molecular biology, was an important site for this discussion in the middle third of the twentieth century (Rasmussen 1997).

Generally speaking, the difference between machine and organism as vehicles for grasping the workings of reality in its entirety has turned on whether intuitions about human craftsmanship provide good models for understanding how things are generally brought into being, even when humans are not implicated. Mechanists trust such intuitions, organicists do not. What accounts for the difference? The historically clearest answer is that mechanists take literally the Biblical claim that humanity's distinctiveness lies in our having been "created in the image and likeness of God". In that case, human creations with at least a certain measure of autonomy in their *modus operandi*, including freestanding machines (the analogues to ourselves, from God's point of view), are reasonably seen as descendants of this original divine creativity. This is how the Scientific Revolution's signature love affair with the "mechanical world view" arose out of various forms of Christian dissent—usually but not always Protestant—from the Church of Rome. However, within two centuries, this vision would be taken to grotesque Gothic proportions in Mary Shelley's *Frankenstein* (Ball 2011).

In short, mechanists hold that humans differ only by degree from God, who is regarded (flatteringly) as The Ultimate Engineer, whereas organicists claim that there is a difference in kind between divine and human creation that will always elude mechanistic models. This elusive difference then serves to distinguish "natural" from "artificial" forms of life, such that life always ends up escaping artifice. In a more theological vein, "nature" also stands for the ineffable, ungraspable character of the deity, even beyond the mightiest of human efforts. Although the theology may have receded over the last two centuries, the persistent openness to the "mystery" of those features that have historically connected the human to the divine—most notably "consciousness"—have continued to motivate the turn to organicism, which nowadays travels under the rubric of "vitalism".

Yet to anyone with a robust sense of history, these appeals to mystery are simply periodic moments of exhaustion of the human will. They do not so much mark a failure of the human imagination as a much more personal failure to take our own imaginings seriously, which is to say, literally. As John the Evangelist would put it, we fail to "make the word flesh". To be sure, at any given moment, there are all sorts of reasons why it might be in our interest *not* to think that we could have the sort of God-like powers implied of the mechanical world view—namely, that if we think it, therefore it should be enacted. Most of these reservations can be classified as "risks", even though that term, strictly speaking, implies that we can calculate outcomes because the imagined future is sufficiently like the past for us to anticipate the likely harms. (To be sure, a confidence in such temporal continuity is itself a very *big* assumption.) Nevertheless, seen from the view of history, most of what we think—however horrible it may appear at the moment of conception—does end up being realized, even if our initial conceptions rarely do full justice to what has come about in our name. To put it glibly, even if the extreme positive and negative scenarios implied by our thoughts rarely, if ever, happen, almost everything else in between does. On this basis, organicism is always likely to prove to be a rear guard position in the face of increasingly sophisticated versions of mechanism that would blur the boundary between *human*, *animal* and *machine*.

Here I allude to the idea of a "cybernetic organism", or "cyborg", introduced by the US mathematician Wiener (1961) at the dawn of the Cold War as the central topic of a book that treated all three entities as variations on a common theme. As Wiener saw it, humans were the most cyborg-like of the three highlighted entities because we exhibit the greatest sense of self-governance, or autonomy. For Wiener, this meant that we do not merely "adapt" to the environment in the weak sense of mirroring it; rather, we make the environment part of ourselves—often with temporary losses—so that it responds either neutrally or it favorably changes in response to our will. Weaker versions of this idea—which retain a sense of what is inside and outside oneself—include "smart environments" and "extended phenotype", while stronger versions encompass "autopoiesis", a literal incorporation of the alien into oneself, very much in the spirit of a Hegelian dialectic. Biomedical scientists will recognize this as a generalization of the concept of *immunity*, itself a physical realization of Friedrich Nietzsche's gnomic maxim, "What doesn't kill me, makes me stronger", a policy that has entered modern folklore through the character of Superman, who acquires his special powers after exposure to radiation.

The prospect of Superman—be it in its Zarathustrian or DC Comic form—begs the question of such a being's transgressive nature: Is it the sheer monstrous hybridity of Superman himself or the fact that his existence points to a continuity between the human and the divine? While the former certainly feeds the popular imagination, the latter points to its ultimate source—namely, that "being" means the same whether one is speaking of the divine or the human—or perhaps even the animate and the inanimate. In modern times, this sentiment is probably most familiar, albeit expressed not quite as originally intended, from the works of Martin Heidegger. Indeed, his own version hides his debt to the medieval master of transgressive semantics (aka "univocal predication", to be explained below), the great scholastic opponent of Thomas Aquinas, John Duns Scotus (1266–1308), on whom Heidegger based his *Habilitationsschrift*, the second dissertation required of German academics for professorial eligibility (Tonner 2008). As we shall see, Scotus introduced a set of general considerations, which when taken together offer the prospect of our developing into some higher state of being, as well as a realization that this higher state may be quite different from our current state of being. In short, the Scotist perspective invites the thought that we are inherently self-transformative beings and hence not tied down to habits of the past. In this capacity, the human may beget the transhuman. But before getting to Scotus' own seminal role, let us start with Heidegger's twist.

In relatively accessible form, Heidegger (2014) proposed in *Introduction to Metaphysics* that humanity is constituted by the temporality of being, which is bounded, on the one hand, by one's being in each moment as it happens and, on the other, the standpoint of eternity, in which one is equidistant from all possible times. Any deity that might exist would inhabit the latter extreme, while the continuum itself can be understood as consisting in increasing levels of abstraction from humanity's intrinsic temporality. In this context, science and technology, understood as abstracting practices that remove one increasingly from the flow of being can be seen as "de-humanizing" precisely as it tries to simulate the standpoint of eternity. This certainly seems to be how Heidegger and most of his existentialist and phenomenological followers have interpreted the situation, which reflects the post-theistic bent of Heidegger's own thinking, which (*contra*, say, Leibniz and Hegel) denies any sense in which becoming "less human" could mean becoming "more divine". As for Scotus himself, the matter is much more ambiguous, given the strictness with which he interprets "being" as having the same meaning (hence "univocal predication"), regardless of what is being talked about. Thus, Scotus defended a version of the "ontological argument" for God's existence, according to which God is defined simply as a being greater than which cannot be conceived. Implied here, once again, is that the deity lies at one extreme of a continuum of being—or "chain of beings", each separated from each other by degree, not kind.

It would be difficult to underestimate the significance of Scotus' move as an incentive to the theological transgressions associated with the Scientific Revolution and, more recently, transhumanism (Fuller 2011, Chap. 2). Specifically, it undermined the appeal to "analogy" that was characteristic of scholastic interpretations of the Bible in Scotus' day—and certainly upheld by Aquinas. That appeal stressed the *negative* side of analogy: Namely, the extent to which Biblical accounts might fall short of reality, given the imperfect natures of its human recorders. At one level, the emphasis on negative analogy looked liberal, especially against the backdrop of Islam's claims for the Qur'an's infallibility as delivered directly by God to Muhammad. It also prevented a present-focused, flat-footed version of literalism of the sort that Augustine had warned against in his own commentary on Genesis. However, the stress on analogy's epistemic shortfalls kept the true/false distinction indefinitely fuzzy, which was resolved in particular cases at the discretion of prescribed clerical authorities—what, in natural law theory, is still called "subsidiarity".

Absent from this superficially tolerant approach to Biblical interpretation was any clear fact-fiction distinction in the modern sense. Modernization effectively required a sense of what is clearly "true" (fact) and "false" (fiction) *for all* at a given moment, while at the same time leaving open the prospect that a proposition might migrate from one side of the divide to the other—say, through the *positive* development of a suggestive analogy. For example, even though not every aspect of human existence naturally lends itself to a mechanistic interpretation, nevertheless the very attempt to "reduce" in point-for-point detail the machine-like aspects of human physiology clearly served to advance the frontiers of knowledge in the history of medicine at various moments (e.g., William Harvey, Claude Bernard, Walter Cannon).

It was just this openness to the exploration of analogical relations—as "models", as we would now say—that Scotus's univocal approach to "being" allowed. Practically speaking, it meant that people were afforded the cognitive licence to leverage their vivid and coherent ideas—even if they have no explicit precedent in reality—as a rational basis for action. Thus, one should not simply succumb to the impulse to dismiss those ideas as mere phantasms. At the same time, one would need to take personal responsibility for whatever ideas they decided to act upon. This is the basis of the modern epistemological idea of "belief" as the expression of a private state of mind for which one is willing to be held publicly accountable. In this context, science provides a relatively safe haven by expressing such ideas as hypotheses, which incur no personal liability, just as long as the proposer submits to the public trial known as an "experiment" (Fuller 2007, Chap. 3). In triggering this line of thought, Scotus broke with Aristotle—and Aquinas—in distinguishing what is logically possible and what is empirically probable, the former providing a conceptual space for what the modern era would recognize as the utopian imagination, whereby something that had been previously "false" (in the sense of "unrealized") may subsequently become "true" (as realized). By contrast, the pre-Scotist imagination was captive to the still commonsensical tendency for people to say "possible" when they mean "probable", and "impossible" when they mean "improbable". By enforcing a stricter form of modal discourse, Scotus opened up humanity's epistemic horizons, which the Scientific Revolution fully exploited, the formal legacy of which is the breakdown of the concept of possibility into "degrees of probability".

Moreover, Scotus took seriously that if God exists, we should be able to say something meaningful about this being's nature, which means that the deity should be discussable in the language we normally use. Indeed, it is a Biblical requirement, given our supposed creation in the image and likeness of this deity, whose privileged mode of communication is language. Thus, God does not simply imprint his will on us—and here Muslims depart from Jews and Christians, at least with regard to the delivery of the Qur'an to the otherwise illiterate Muhammad. Rather, we are God's interlocutors, in a manner that was updated in the twentieth century by Martin Buber's "I and Thou". Not surprisingly then, Scotus talks more about God than most other scholastics—and certainly much more than Aquinas. He rejects the "analogical" reading of the Bible, which makes it easy to dismiss talk of God as "all powerful" as using "powerful" in a non-literal fashion. On the contrary, it is just this literalness (i.e., "univocity") in the meaning of "powerful" when applied to both ourselves and God that Scotus insisted upon, as an implication of our own being (as *imago dei*). Language is not something we invented to understand God; rather, it marks our relatively poor—yet improvable—expression of our own God-like character, who after all creates through the word. In this regard, it is disappointing that Biblical literalists and scientific operationalists fail to see the similarity of their self-appointed epistemic remit, courtesy of two great legalistic minds, John Calvin and Francis Bacon. Both took univocity with the sort of deadly seriousness that makes them part of the modern secular world view, one that gives enormous significance to "creation by the word", be it through the laws of man or the laws of nature (Fuller 2010a, Chap. 5).

Last but not least, if a predicate attributed to God and humans means the same in both cases, then the difference between the two can be discussed in terms of a continuum, such that we can literally compare ourselves to God. In that case, the prior existence of humanity's Original Sin provides a motivation for humans to move along the continuum from "less" to "more" in terms of these divine attributes, though in the end God judges whether we have truly made the move, however good our intentions. Moreover, Scotus was aware that all the perfections that are convergent in God (i.e., his "all-powerful", "all-good", "all-knowing", etc. nature) exist to varying degrees—in different states of imperfection—as human virtues (e.g., the intelligent people are not necessarily the good ones). Indeed, God may be defined as the limiting case of our imagination, as a being who jointly maximizes all virtues simultaneously. However, the Franciscan order to which Scotus belonged generally believed that this imaginative shortfall need not remain mysterious but could be remedied through proper mental training, as in *The Mind's Journey to God* by the great thirteenth century Minister General of the Order, Bonaventure (Fuller 2015, Chap. 2).

However, one consequence of such training is that virtues that appear separately and perhaps somewhat at odds in humans may start to look different as we come closer—"evolve", if you will—to their joint instantiation, as they exist in God. In other words, the sensibility associated with, say, "being good" may change as we get to very extreme cases that approach divinity. After all, the morality of the saint or the martyr is not that of the father or mother, the intelligence of the scientist is not that of the carpenter, and so forth. Thus, while we may be doing the best we can in light of the information at our disposal and our capacity to anticipate consequences, we should expect that as our cognitive horizons expand, the moral character of our actions might change as well. The somewhat problematic standing of *compassion* in Christian theology is worth considering in this context. Although there is no denying Jesus' compassion, many theologians have seen this virtue as a feature of his human—but not divine—nature, as it involves forming a bond based on the target creature's current physical condition. In contrast, God understands and values his creatures through the lens of their ultimate cosmic significance, which while generally positive is not necessarily related to how they experience their own immediate circumstances. The popular, secular version of this sensibility appears in the rather detached and somewhat enigmatic demeanor of "superior beings" in science fiction.

Perhaps more consequentially, the Scotist image of the divine convergence of virtues resurrects the idea of a being just short of divine—Plato's philosopher-king—who merges the worlds of science and politics in his style of rule. In the wake of the Second World War, and against the spirit of the times, Alexandre Kojève famously (albeit unsuccessfully) tried to persuade Strauss (2000) of the prospect of rulers in the future who would dialectically overcome both the abstractness of science and the short-term horizons of politics, as well as many of the other contradictions that define modern ethics and epistemology. Even if this quest to realize the philosopher-king fails to get us closer to God, it remains intriguing as both an intellectual and a political project (Fuller 2011, Chap. 5). As a concluding case in point, consider the tremendous amount of philosophical ink spilled on the supposed distinction between "utilitarian" and "deontological" approaches to normativity. The utilitarian proportions the desirability of ends in terms of the availability of means, so as to produce the greatest good for the greatest number, whereas the deontologist pursues an end for its own sake, regardless of cost and consequence. The former position derives its plausibility from the need to make trade-offs under conditions of scarcity, the latter from the need to act in spite of profound ignorance of the empirical upshot of our decisions. Both try to generate virtue out of human liabilities from which God does not suffer. In God these seemingly contradictory ethical horizons are integrated into a coherent whole. And in the person of the philosopher-king, this "coherent whole" acquires a concreteness that can arguably function as a principle of authority in society, insofar as each subject can see him- or herself as the philosopher-king sees them. This would amount to a political version of the categorical imperative, the content of which would be one's social utility function. In this respect, the identities of the king and subject would be blurred.

References

Ball P (2011) Unnatural: the heretical idea of making people. Bodley Head, London

Bloom H (1973) The anxiety of influence. Oxford University Press, Oxford

Blumenberg H (1985) Work on myth. MIT Press, Cambridge

Fuller S (1988) Sophist vs. skeptic: two paradigms of intentional transaction. In: Otto H, Tuedio J (eds) Perspectives on mind. Kluwer, Dordrecht, pp 199–208, 389–390

Fuller S (2005) Philosophy taken seriously but without self-loathing: a response to Harpine. Philos Rhetoric 38:72–81

Fuller S (2007) New frontiers in science and technology studies. Polity, Cambridge

Fuller S (2010a) Science: the art of living. Acumen, Durham

Fuller S (2010b) Thinking the unthinkable as a radical scientific project. Critical Review 22 (4):397–413

Fuller S (2011) Humanity 2.0: what it means to be human past, present and future. Palgrave, London

Fuller S (2015) Knowledge: the philosophical quest in history. Routledge, London

Fuller S, Lipinska V (2014) The proactionary imperative: a foundation for transhumanism. Palgrave, London

Galison P, Stump D (eds) (1996) The disunity of science. Stanford University Press, Palo Alto

Heidegger M (2014) Introduction to metaphysics, 2nd edn. Yale University Press, New Haven

Hesse M (1963) Models and analogies in science. University of Notre Dame Press, South Bend

Hillar M (2009) The polish socinians: contribution to freedom of conscience and the American Constitution. Dialogue and Universalism 19(3–5):45–75

Passmore J (1970) The perfectibility of man. Duckworth, London

Popper K (1972) Objective knowledge. Oxford University Press, Oxford

Rasmussen N (1997) The mid-century biophysics bubble: Hiroshima and the biological revolution in America, revisited. Hist Sci 35:245–291

Strauss L (2000) On tyranny. University of Chicago Press, Chicago

Tetlock P (2003) Thinking the unthinkable: sacred values and taboo cognitions. Trends Cogn Sci 7(7):320–324

Tonner P (2008) *Haecceitas* and the question of being: Heidegger and Duns Scotus. Kritike 2(2):146–154

van der Laan JM (2007) Seeking meaning for Goethe's Faust. Continuum, London

van der Laan JM (2010) Lessing's "lost" faust and faustus socinus. In: Schade R (ed) Lessing yearbook 2008/2009. Wallstein Verlag, Göttingen, pp 53–67

Wallace D (1984) Socinianism, justification by faith and the sources of John Locke's The Reasonableness of Christianity. J Hist Ideas 45(1):49–66

Weir T (ed) (2012) Monism: science, philosophy, religion, and the history of a worldview. Palgrave, London

Wiener N (1961) Cybernetics: or control and communication in the animal and the machine, 2nd edn. MIT Press, Cambridge

Williamson O (1975) Markets and hierarchies. Free Press, New York

Wolbring G (2008) The politics of ableism. Development 51:251–258

Yates F (1964) Giordano Bruno and the Hermetic tradition. Routledge & Kegan Paul, London

Index

© Springer Fachmedien Wiesbaden GmbH 2016

B.-J. Krings et al. (eds.), *Scientific Knowledge and the Transgression
of Boundaries,* Technikzukünfte, Wissenschaft und Gesellschaft / Futures
of Technology, Science and Society, DOI 10.1007/978-3-658-14449-4

Printed in the United States
By Bookmasters